Praise for
A Wilder Time

"While conveying the geological hypotheses, techniques of data collection, and adventures of his expeditions to Greenland with his two Danish colleagues, William E. Glassley also brings startling sensory precision to his descriptions. The velvety feeling of moss, the taste of lichen, the alternating rhythms of terror and fluidity in schools of fish through which a predatory sculpin cruises—such experiences bring what might have seemed a stark world of rock and ice alive. This delicacy of perception is the vehicle through which not only the scientific quest but also the profound mystery of our living Earth saturates this memorable book."

—**John Elder**, coeditor of *The Norton Book of Nature Writing* and author of *Picking Up the Flute*

"*A Wilder Time* is a wonderful mix of science and poetry. It delves into the kind of spiritual effect that wilderness has on those privileged to work in it and how it changes the way we experience and understand our surroundings and our lives. The science, including the geological controversy at the heart of the book, is lucidly explained, and readers will be absorbed by the story Glassley tells as well as his many vividly described encounters with nature. Next time someone asks me why I am a geologist, I will just hand them this book."

—**William L. Griffin**, professor of geology at Macquarie University

"While conducting research probing deep time and the origin of continents, Glassley discovered a further source of fascination: the Arctic wilderness of Greenland. In *A Wilder Time*, he shares his encounters with unvarnished nature still free—for now—from the corruptions and constructs of human settlement. With openness, clarity, and a keen eye for detail, he weaves adventure, research, astonished awe, and thoughtful reflection into an absorbing account of his sojourns."
—**Martha Hickman Hild**, author of *Geology of Newfoundland: Field Guide*

"As geologists, we may be rational scientists, but expeditions to remote places touch something deep in us that moves us to also be poets. Glassley has turned his experiences in Greenland into searingly beautiful descriptions of a wild landscape and the ways in which that landscape moves and changes him. Every sentence is evocative, connoting curiosity, awe, and respect in equal measure. *A Wilder Time* is a paean on the importance of wilderness to the human spirit and a saddening reminder of what we lose when we divorce ourselves from contact with wild places. Glassley's voice will stay with me the way the works of Loren Eiseley, Edward Abbey, Rachel Carson, and Aldo Leopold have stayed with me over the decades."
—**Jane Selverstone**, professor emerita in the Department of Earth & Planetary Sciences at the University of New Mexico

A WILDER TIME

A WILDER TIME

Notes from a Geologist at the Edge
of the Greenland Ice

William E. Glassley

BELLEVUE LITERARY PRESS
New York

Frontispiece: *John and Kai walking toward the ice edge*

First published in the United States in 2018
by Bellevue Literary Press, New York

For information, contact:
Bellevue Literary Press
NYU School of Medicine
550 First Avenue
OBV A612
New York, NY 10016

Library of Congress Cataloging-in-Publication Data
Names: Glassley, William E.
Title: A wilder time : notes from a geologist at the edge of the Greenland ice. /
William E. Glassley.
Description: First edition. | New York : Bellevue Literary Press, 2018.
| Includes bibliographical references.
Identifiers: LCCN 2017005087 (print) | LCCN 2017006260 (ebook) |
ISBN 9781942658344 (pbk.) | ISBN 9781942658351 (e-book)
Subjects: LCSH: Geology—Greenland. | Geology—Research—Greenland. | Climatic
changes—Research—Greenland. | Climatic changes—History. | Ice cores—Greenland.
Classification: LCC QE70 .G53 2018 (print) | LCC QE70 (ebook) | DDC 559.8/2—dc23
LC record available at https://lccn.loc.gov/2017005087

Bellevue Literary Press would like to thank all its generous donors—
individuals and foundations—for their support.

 This publication is made possible by the New York
State Council on the Arts with the support of Governor
NYSCA Andrew Cuomo and the New York State Legislature.

 National Endowment for the Arts This project is supported in part
by an award from the National
ART WORKS. arts.gov Endowment for the Arts.

Book design and composition by Mulberry Tree Press, Inc.

Manufactured in the United States of America
First Edition

1 3 5 7 9 8 6 4 2

paperback ISBN: 978-1-942658-34-4
ebook ISBN: 978-1-942658-35-1

To Kai Sørensen and John Korstgård,
whose friendships, hearts, and souls
made Team Alpha happen,

and to Nina,
who forced me to accept the moment

Either everything's sublime, or nothing is.

—KATHERINE LARSON

You didn't come *into* this world.
You came *out* of it, like a wave from the ocean.
You are not a stranger here.

—ALAN WATTS

Contents

List of Maps and Figures

Preface

DESTINATIONS, WHETHER NEW OR OLD, are expectations shrouded in imagined landscapes. We set off with ideas for adventures that we hope might materialize and we imagine pathways to things we fear but secretly wish to confront. We think of our destination as an end point to a journey, but its reality is seldom that. Destinations may also evolve into portals that devour expectations, immersing us in the inconceivable. So it is for me when I travel into the Greenland wilderness.

FOR A GEOLOGIST, GREENLAND IS A DREAM. The ice, receding faster than plants can take hold, exposes in its retreat the polished bedrock floor it has ridden on for millennia. Glistening in the sun, emphatic in its insistence for attention, an unexpected artistry offers itself for inspection.

That rock can flow always astonishes, but revealed in those outcrops are patterns that imagination could never conjure, proving beyond doubt that the continental heart is barely less fluid than water. Layer upon layer, some a fraction of an inch thick, some thicker than houses, colored in a palette of earth tones and off-whites, greens and blue-blacks and reds, fold back on one another, pinch and swell, stretch to paper thinness, then thicken again, telling stories we ache to know but can barely read.

I go to Greenland with two Danish geologists, Kai Sørensen and John Korstgård, to unravel these mysteries. For weeks at a time, camping in one of the world's greatest untouched wilderness expanses, we wander through a twenty-thousand-square-mile landscape, crawling over outcrops on hands and knees, struggling to piece together fragmented clues to what the story line may be. It is the ultimate in forensic science, cobbling from a hundred different techniques, technologies, and fragmented logical arguments a consistent tale that encompasses nearly the entire history of nonhuman Earth.

Our research, and that of colleagues stretching back to the 1940s, has provided only the barest outline of that history. We have established little beyond that it is a mystery involving life and rock and the symbiosis they have woven. If a book were the analogy, the covers would be mainly complete but the ink of the chapters nearly faded.

That so little has been accomplished should not be surprising. The region is above the Arctic Circle, so daylight and sufficiently warm temperatures to allow camping are available only a few months during the year. The remoteness of the area, requiring special arrangements for transportation into and out of the wilderness, challenges logistical efforts. It is a vast terrain of unexplored landscape; only a few details are well established.

What has been exposed thus far is a tantalizing

mystery. Preserved in the bedrock are vague suggestions that multiple mountain-building episodes occurred there sometime between two and three-and-a-half billion years ago. The most recent of those events may have been so massive, it would have foreshadowed the Himalayas. There is evidence of movement along enormous faults; of volcanic systems that would rival the Andes; of ocean basins the size of the Atlantic. All are now vanished, swallowed in the onward rush of Earth's evolution. The observations supporting these notions are few, the data difficult to interpret.

Compounding the challenge of this research has been uncertainty about the fundamental assumptions upon which the science is based. All geological studies dealing with present-day processes on Earth are grounded in plate tectonics. Plate tectonics defines the Earth as a dynamic planet in which heat from the deep interior powers the slow migration across the surface of twelve plates of ocean and continental crust. Mountains form where plates collide, and crust forms where plates separate—the consistency of the process of crust creation and destruction fulfills the requirements of a self-contained system, a zero-sum game. Recognized and accepted evidence for the persistent operation of this process extends 900 million years into the past. Beyond that time, the evidence is equivocal and energetically debated. Since the rocks in Greenland are much older than that, we are left uncertain as to how to interpret what we see and what the motivating forces would have been.

The rocks we work on are from a transition period. Life, though soft-bellied and delicate, has been the most powerful chemical agent on Earth. The atmosphere of our planet is a product of its breathing, the composition of oceans and rivers a consequence of its metabolism. Even the continents are its product—over 3,800 million years ago, the remnant structures of photosynthesis, mixed into the mantle, encouraged melting that oozed from that deep interior, coalescing to become the landmasses we walk on.* Was that when plate tectonics began, or was plate tectonics some later phenomenon predated by an energetic process we do not know? The rocks we collect and study preserve the answer to that question.

WE CONDUCT OUR STUDIES IN that little-known fringe of land that extends over a hundred miles west from the edge of the Greenland ice sheet. Although our scientific interests are purely academic, the experiences we have lived through are almost mystical. We camp for weeks at a time in one of the world's largest unbroken wilderness realms. Utterly alone, voluntarily isolated from the rest of humanity, we walk and sail without resistance through a world that, for the most part, has never experienced the presence of a human being. We sample and photograph and measure the incomprehensibly old bedrock that preserves

---

* M. Rosing, et al. 2006. The rise of continents—an essay on the geologic consequences of photosynthesis. *Palaeogeography, Palaeoclimatology, Palaeoecology* 232:99–113.

nearly the entire history of the planet. Although harsh and unforgiving, that wild surface is engulfed in beauty, revealing an exuberantly evolving world.

Wandering and sailing from outcrop to outcrop, immersed in the grandeur of that wilderness forces daily life to become a practice in humility. Time fractures, languishing in some backwater of perception. Viewing ice, somnambulant fjord waters, rocky defiles, and tundra plains becomes a repeated experience of confronting the incomprehensible, each thing expressing a subtle essence of existence that can be known only by being present. The gulf that exists between the prejudiced expectations derived from urban life and the bedrock purity of that wild landscape is nearly unbridgeable. The feeling that I have become alienated from, and ignorant of, such purity is inescapable and devastating.

I now understand that wilderness is as much story as it is place. Untouched lands provide inspiration and nourish our imagination with mysteries and connections impossible to conceive anywhere else. The depth of their richness, the complexity of their structure are beyond common experience. Wilderness is the primordial heart of what we conceive of as soul, and as a consequence, it must be accepted as a version of home. For me, Greenland has been the landscape that embodies that lesson. Ironically, perhaps, it was the pursuit of quantitative, objective observations that exposed the emotional truths contained in wild places.

THE WORD *wilderness* derives from the Old English word *wildēornes,** meaning "the place where only wild animals live." Implicitly, that word also defines the place where human existence is inherently a struggle. It is the land in which it is not easy to settle, to farm, raise families, or enjoy an evening with friends. Wild places where only animals live are the frontiers, the lands through which humans may have wandered, but within which living is likely to fail. Wilderness is not welcoming. It is the place where humans may be prey.

Once, wilderness was everywhere, an ambience for wandering humanity from the time of our origins. Many languages have no word for wilderness because it simply was the context of existence—naming it was unnecessary. Now, we are no longer wanderers—over the last thousand years we have started naming wilderness because it is almost gone. We have flowed over the surface of the planet like a massive tsunami, inundating the world with more and more beings while pushing to the fringes any possibility of the experience of deep wilderness. Within thirty-five years, the population of the planet will grow from over seven billion people to more than ten. As it does so, wilderness will passively retreat, taking with it the only opportunity we have to know our true origins. Without immediate contact with the offerings of wild

---

* The pronunciation of the word is unclear, given that the language has not been commonly spoken for nearly 900 years. It is believed by some that it would be recognizable if spoken to modern English speakers.

lands, we lose the world that is the foil for humanity. Tragically, we barely notice, even when it is obvious. I give testimony to this fact—I was an unintentional witness to a relict of this loss.

One evening, while Kai cooked and John refined his notes, I walked along the shore north of our little camp, seeking a quiet place to reflect on the day. I hiked over a low ridge and found an unexpected, modest bay. The tide was low; small waves lightly lapped far out at its mouth. I went down to the narrow beach, where the most minuscule of slow-moving ripples, descendants of the little waves farther out, migrated across the watery membrane that rested on the succulent bay muds. Icebergs floated in the fjord waters farther out. The pinkish gray light of the dappled cloud underbellies reflected off the water skin that barely submerged the sediments. What little drama there was, the mind created in imagined eyes and stalking creatures hidden in the black shadows cast by hundreds of boulders—a few inches to several feet in diameter—that were scattered on the exposed floor of the bay. For many long moments, I calmly drank in the lush scene. But slowly, something incongruous began to disturb the moment—something below the surface of what I chose to see. As I focused on the boulders, I saw one that, oddly, had a small tundra hummock, delicately balanced, resting on its top. The small layer, a few feet thick, flat-topped, with tall grasses growing from it, looked as though someone had delicately placed it there. Trying to make sense of it, I then noticed that every boulder above a certain size had an exact duplicate of that little tundra mound. The flat top of every tundra cap was at exactly the same elevation.

Stunned, I realized each tuft was an erosional remnant of a tundra plain that had, in the very recent past, reached as far out as the edge of the bay. But rising sea level had eaten away at the delicate vestiges of the plant remains and the boundary that once defined a land—tide harmony. The edge of wilderness, offering little resistance, was silently retreating into a new future we unknowingly are shaping.

When wilderness is gone, even that which is responding naturally to climate change forces, all that will remain are memories and impressions of its textures and forms, its silences and screams, its smells and tastes. We will have lost the only reference point we have for the significance of mind in the universe.

As days went by while I camped with John and Kai in the wilderness of West Greenland, the noise of cities receded into dim memories, and self became a part of the landscape. Boundaries dissolved between what is external and what is internal to the soul. Who and what we individually are became a shared question with how Earth had evolved. What we scientists went there to study and resolve melted into the background of an incandescent experience of place.

A WILDER TIME

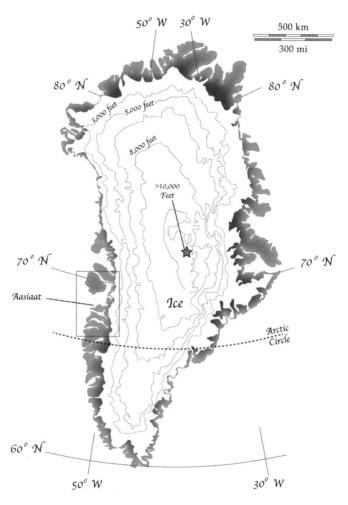

*Greenland, ice thickness and land areas (dark gray).
The boxed region encloses our research area.*

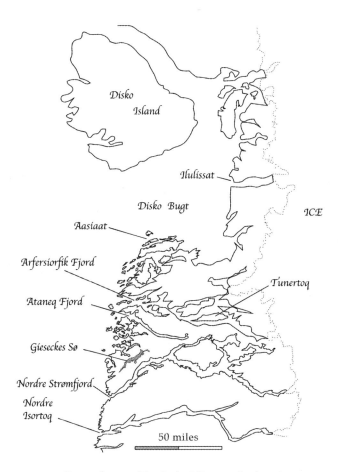

Research area. The dashed line marks the edge of the inland ice sheet.

Introduction

ONE OF THE MOST EXTENSIVE, continuous wilderness regions on Earth, Greenland remains largely submerged by ice. In the area not ice-covered, the landscape materializes as experience, not place. Boundaries, whether real or imagined, named or anonymous, dissolve into opportunities. Senses become remarkably acute, sharpened by the raw purity of what it means to be wild. Greenland is a place of surfaces so rich with history that simply setting foot on them seems to clarify reality.

The objective meaning of Greenland, expressed as simple facts, deserves consideration. That land of rock-fringed ice, if laid onto western North America, would extend beyond the northern and southern borders of the United States, and stretch from San Francisco nearly to Denver. More than 80 percent is buried under the only permanent ice sheet in the northern hemisphere. At its thickest, the ice is more than ten thousand feet deep and holds more than 10 percent of the world's freshwater. The summit of the ice cap is over twelve thousand feet above sea level.

More than half of Greenland extends above the Arctic Circle. It was the last settled major landmass on the planet, its first population reaching it about 4,500 years ago. It holds the distinction of being the least-populated region in the world, and is the only nation listed in the World Bank database with a value of zero people per square kilometer (the database presents all statistics as whole numbers). That same metric for the United States is 35 people per square kilometer; for the United Kingdom it is 265. Most of its fewer than sixty thousand permanent residents identify as members of the Inuit culture. The largest town is Nuuk, with 16,500 people. There are only seventy-eight towns, villages, communities, and settlements on the entire island. A number of them have fewer than fifty inhabitants. The Inuit culture identifies their country as Kalaallit Nunaat.

Greenland's culture is steeped in its fishing and hunting traditions, sustainably practiced for hundreds of years. Seal and reindeer are essential staples, providing nourishment and materials for clothing and limited commerce, the hunting done by individuals as part of a subsistence lifestyle. The art, photography, literature, and inherited myths of its indigenous Inuit peoples quietly offer perspective on their home and traditional practices. But in the absence of any significant capitalized trade, few nonnative people have access to it, or can see how it is changing.

The ripple effect of distant decisions made by countries navigating the complex interactions of economics, morality, and the wild world extends even to a place as remote as Greenland. In 1983, in reaction to the attention given to the brutal commercial harvesting of baby seals in Canada, a ban on sealskin trading was imposed by the European Economic Community, followed in 2009 by a European Union ban on trade in seal products. The consequences were far-reaching, some of which were unintended. A loss of income from the sale of sealskins and other products devastated Greenland's Inuit hunting culture. The extinction of the seal market diminished seal hunting, causing an explosive growth in seal populations. With a rapid expansion in the number of fish predators, fish populations consequently declined, impacting that component of their subsistence lifestyle, as well. Even with very recent modifications to the ban—those allowing Inuit cultures to pursue sustainable seal harvesting— the impact on income has been significant. Today, about 60 percent of Greenland's economy is supported by an annual block grant from the Kingdom of Denmark, of which it is an independent member. Greenland remains a country struggling to return to a sustainable existence, but now with the added complexity of a rapidly changing climate, the challenge is formidable.

WHAT FOLLOWS ARE MY EXPERIENCES of Greenland's surfaces from six expeditions. The story unfolds in three parts, each part containing the suite of formative sensory experiences that shifted my perception. "Fractionation" documents the deconstruction of expectations, relating experiences that exposed the depth of my ignorance about knowing place. "Consolidation" describes the process of coming to terms with the reality that, as a product of organic and physical evolution, my ignorance is an integral part of being aware. "Emergence" derives from small epiphanies about our place in existence, what we can know of the world and what we cannot.

That we have a place in this world implies responsibilities, but it does not signify meaning. The majestic power of wilderness is its ability to convey that seeming contradiction through the overwhelming beauty of evolution's carelessness. That we have an impact on its unfolding is revealed in the reconstruction wilderness imposes upon itself when confronted by changes in climate that mankind has induced and to which wilderness must respond.

The book is not chronological. Experiences that change perception accumulate in odd ways that are personal and often not initially understood. The reconstruction of a new *way* of seeing is piecemeal. Each insight or shifted perception fills a space in a timeless tapestry that will never be completed.

Wilderness speaks with unmitigated honesty. Every belief and imagining that we bring with us as we enter such spaces also reflect back to us, but in a form that can be difficult to recognize. My hope in writing this book is that the value of truly pristine wilderness, as a place from which to sense how we each fit within the grand unfolding universe, will inspire its preservation. If we lose wilderness, finding our roots, personally and as a species, will be virtually impossible.

IMPRESSIONS I

Beauty itself is but the sensible image
of the infinite.
—George Bancroft

ALL THAT WE SEE IS SURFACE. What we perceive as experience derives from light reflected, a product of events that have flowed to the present and become, in a moment, a shape seen. Life teaches us to extract texture and form, weight and warmth from that impression.

But what is it that silently rests below that cosmic skin, composing the thing we sense? We reach toward the stars to understand why the sun rises, why winter comes, why we must die. And yet, what we find in each answer and insight is a deeper question, an underlying complex of mysteries that serve only to feed our imagination. With these fragments, we construct a body of knowledge about the components of our world, each of us building a unique framework that becomes the context for our individual lives, the thing upon which we hang notions of meaning.

Through this process, we have come to realize that life is an unstoppable force that endlessly evolves, eventually achieving the emergence of mind from stardust and time. And yet, despite the stupefying significance of this revelation, we also see that, from a cosmic perspective, we are a trivial event. We are a speck on a flowing river of entropy that still gushes from an unfathomable beginning nearly fourteen billion years ago. We're enthralled by a story we suspect the stars possess, but we remain unable to grasp its outline. We wander over landscapes, looking for histories the stones sequester, hoping there will be in them a flicker of an insight that will expose something worth cherishing.

FRACTIONATION

One thing had impressed us deeply on this little voyage: the great world dropped away very quickly. We lost the fear and fierceness and contagion of war and economic uncertainty. The matters of great importance that we had left were not important. There must be an infective quality in these things. We had lost the virus, or it had been eaten by the anti-bodies of quiet. Our pace had slowed greatly; the hundred thousand small reactions of our daily world were reduced to very few.

—John Steinbeck

Silence

THE BOAT THAT BROUGHT US into the field was a fishing trawler chartered by the Geological Survey of Denmark and Greenland. It had a baby blue hull, a weathered, varnished wheelhouse that two people could cram into, and a worn wooden deck, onto which we had piled the backpacks, crates, tents, a few bags of fresh food, and other gear meant to sustain our little expedition. John, Kai, and I met the boat in Aasiaat, West Greenland, on the southern edge of Disko Bugt. Aasiaat is one of the largest towns in Greenland, with a population of just over 3,100 people. Walking through every street, passing every house, would take a few hours on a summer afternoon.

Under the watchful eye of Peter, the skipper, we had spent half an hour loading the trawler, securing the gear, and inventorying before setting off into the iceberg-studded waters. The trip would take many hours, so we took turns napping in the tiny forecastle, where two bunks were tightly bolted to the bulkhead. The sound of the sea swishing by could be heard through the hull's three-inch-thick oak planks. I slept for about an hour, then went back on deck to watch the scenery.

The air was still and cool, the water like glass under

an overcast sky. Whales occasionally breached in the distance, feeding on schools of small fish at the surface. We passed by skerries, some with packs of huskies that had been left there for the summer by their masters. The sled dogs were nearly feral.

I leaned against the peeling rail, mesmerized, the *chug-chug-chug* of the two-stroke diesel thumping in the background. I was warmly dressed in a field shirt, sweater, and fleece jacket, a woolen skullcap pulled down to my ears, my body braced against the forty-degree chill.

As the islands passed, the world I was leaving behind tugged with an unexpected angst. I had been anticipating the expedition for months, looking forward to sharing with old friends what I knew would be daily discoveries in a virtually unexplored terrain. But an aching sorrow overwhelmed that excitement—my wife and daughter would not be seen or heard for months, the sweet pleasures of family life erased, the known small comforts of cooking meals together, sharing movies, reading the newspaper, laughing with friends at parties, taking Nina to the bus for school—gone.

My contemplation was broken when the first mate came up and leaned against the rail next to me. His sand-colored hair was matted; his blue eyes blazed in a weather-beaten face. His nose, broad and flat, made it clear he had some history. His English was perfect, but with an accent I didn't expect.

"So, what're you guys doin' up here?" he asked. Despite

the cold, he was dressed in a short-sleeved T-shirt and jeans.

"We're geologists," I said, quickly recovering a semblance of composure. "We're here to study the rocks."

He thought for a moment and then said, "Hmm. Lookin' for gold?"

"No, just interested in the history of the rocks."

He nodded and pursed his lips.

"Why is that interesting?" he asked nonchalantly. He wasn't looking at me—his eyes were on the slowly passing scenery.

I explained that there was some debated evidence that a mountain system about the size of the Himalayas or the Alps had existed there nearly two billion years ago. Now all that was left were cryptic hints preserved in what might have been the deep roots of that old mountain system. After so much time, erosion had brought those potential roots to the surface, where we could study them to see if that story were true.

"Mountains like that here? That's really amazing . . . hard to believe," he said as we both looked out at a rolling landscape that gave no hint that K2s and Eigers and Mount Everests had once soared there.

"Where're you from?" I asked. His Anglo complexion and accent made it obvious he hadn't been born and raised here.

"Sydney. I came here with my girlfriend five years ago. We were just tourists but hung around because it was so beautiful. I ran into Peter a couple of times and got to like him. He's Swedish. Been here twenty-five years. He goes back to visit family in February but has to return here— no place else he can live. Our first year here, we took care of his house when he went back. When he returned, he offered me a job on his boat, and I took it."

He looked out over the water for a while and then said, "I can't go back to Australia. It's too hot." He laughed. Then he got serious.

"I love the life here. It's free and open. There are too many people in other places. . . . People here take care of each other. But they understand that it's what's out there that matters." He waved his hand at the horizon. "There's a peace here, an emptiness that I've never seen anywhere else. . . . I can't give that up now. Neither can my girlfriend. This is home now."

I looked out at the landscape and wondered what he felt as he looked at it. I loved my neighborhood in the San Francisco Bay Area, the streets and cafés and small shops, but that connection obviously paled in comparison with his passionate relationship to place.

For a long while, nothing was said. Then he pushed back from the rail. "I better get back to work. Peter hates it if he's payin' me and I'm not doin' somethin' on the boat. Good luck. I hope you find what you're lookin' for out there." He shook my hand and walked away.

THE JOURNEY THAT HAD LEAD TO THAT MOMENT was a long one, stretching across years and half the globe. I had met Kai Sørensen nearly three decades earlier, in Oslo, Norway. He was from Denmark, escaping a complicated situation involving love and friendship, while simultaneously trying to pursue his scientific career in geological studies. He had come to the research institute where I was to find a mental refuge where he could quietly continue his research and reconstruct his life.

I, too, was seeking change. I had just been through a divorce, started a new relationship, and finished my Ph.D. When the chance to pursue new research directions in Norway was offered to me, I jumped at it, craving a place where I could start over. I knew no one in Oslo, which provided the possibility of a monastic lifestyle, a quiet world where immersion in a science I was just beginning to understand could be an escape from a complex emotional past. Our somewhat similar state of emotional and cultural transience resulted in many discussions, a shared apartment, and a close friendship. Eventually, we were joined by a third, Julian Pearce, whose life path mirrored ours in many ways. We became an odd household of foreign friends. Each morning, we rode the bus to the research institute, ate lunch at the communal table of geologists on the third floor, and rode back at night to take turns making dinner. In the evening, we played hearts, which I nearly always lost, listened to *Cabaret* and *Jesus Christ Superstar* on Kai's stereo, and sipped coffee enhanced with a shot or two of Linie aquavit. In that temporary setting, we found stability.

THE DESIRE TO CHANGE RESEARCH DIRECTIONS was stimulated by a growing excitement I could not have anticipated when I began pursuing geology. During the first few years of my thesis work on the relatively brief sixty-million-year geological history of the Olympic Peninsula of Washington State, I slowly began to perceive the incomprehensible magnitude and beauty of Earth's evolution. I was overwhelmed by the unstoppable, yet unimaginably slow, dynamism eloquently detailed in the bedrock backbone of landscapes. I became addicted to the thrill of experiencing unseen and unrecognized histories of much more ancient times. The position in Norway provided an opportunity to work on problems more profound than what my thesis research had considered. The work at the research institute in Oslo was a chance to deal with fundamental questions, such as how certain types of rocks exchanged chemical compounds with other rocks when buried tens of miles below the surface. It was an esoteric academic issue, of little interest to any but a handful of other researchers scattered around the world, but it also allowed me an opportunity to delve into something that had global implications, even if on a virtually insignificant scale.

While involved in those studies, Kai would tell me captivating tales of the work he was doing in West Greenland in a terrain of very old rocks with a complex history. The setting, at the edge of the Greenland ice sheet in a place I knew nothing about, deeply intrigued me. He described mysterious patterns in rocks over two billion years old that seemed to record events very much like those happening near the land surface in today's Himalayas or Alps. Those ancient events in Greenland seemed to have taken place many miles below the surface, possibly preserving hints as to what is happening today far below the jagged peaks of those present-day mountains. But there was no obvious plate tectonics context within which to fit those observations—the rocks were too old and too little was known about those ancient times to allow anything other than empty hypothesizing.

His specialty was structural geology, which meant he focused his attention on the shapes, patterns, and orientations of layers in the rocks. He and his colleagues had reached the conclusion that the area was a complex zone where it seemed a continent had literally fractured, with one part slipping past the other for many tens or hundreds of miles shortly after the mountains had formed. It was an area of intense deformation.

I had a background that could complement their structural work, providing details about temperatures and pressures the rocks experienced as they went through that extreme deformation. My expertise was in metamorphic processes, which meant using the minerals in rocks to decipher how hot they had gotten and the paths they had followed deep into the earth and back again. Working in laboratories with microscopes and X-ray spectrometers and electron beams, I could tease from the rocks their journeys through vast times and great distances deep into the earth and back to the surface. Just before returning to the U.S. I convinced him to let me work in the lab on the rocks he had collected, hoping one day it would lead to visiting the place.

Eventually, I became friends with John Korstgård, a colleague of Kai's who also was primarily a structural geologist but who had extensive experience in geochemistry and mineralogy. The three of us made a good team.

After a few years, we obtained funding to travel to Greenland and then worked there together, enjoying our collaboration. For nearly a decade, we pursued common interests, publishing a few papers and giving joint presentations at conferences. But over time, our attentions were distracted by differing career paths and life choices. By the late 1990s, our communication was only occasional, and the work in Greenland a fond memory.

Unexpectedly, Kai got in touch with me in 2000 about plans for a new expedition. At that time, he was involved with the Geological Survey of Denmark and Greenland, which was sponsoring a regional research effort in West Greenland. He asked if I would be interested in joining him and John in new work there. It would be a chance to expand our earlier work into areas we had not been able to explore before because of budget and time constraints. In passing, he also mentioned that there was some controversy about the earlier interpretations he and others had made about the significance of the zone of intense deformation; resolving that controversy would also be part of the effort.

Although I was not directly involved in research in Greenland at that time, I followed the research that was published, simply out of personal interest. I was aware of a few papers that had appeared that presented interpretations of the history that were inconsistent with what I had learned from Kai and John and some of their colleagues, but I had dismissed them. I had assumed those papers were simply offering options for consideration and had not been taken very seriously by the research community. I had no idea that a deeper personal conflict was hidden behind the scenes.

Craving a return to Greenland, and to work closely again with John and Kai, I jumped at the chance to join their expedition. For years, I had been quietly nagged by memories of unanswered questions in the work we had pursued.

Standing at the rail of the boat, watching the skerries float by, I could not have anticipated that we were on the first leg of a journey that would carry us more than fifteen years into the future.

WHEN WE REACHED THE SITE of our planned base camp, the skipper pulled the trawler into a cove and we off-loaded the gear, using a small skiff. It took several trips, but within thirty minutes all the supplies were stacked at the foot of a small bluff on the beach. When we were finished, we shook hands with the mate and the skipper and said our good-byes.

Our campsite sat on a narrow, ragged bench that ran along a stretch of the northern coast of Arfersiorfik Fjord.

We were ten miles west of the inland ice cap, sixty miles from the nearest Inuit settlement, and far enough above the Arctic Circle that the sun would not set for weeks.

A chill evening breeze blew. I turned up the collar of my parka, jammed my hands into its pockets, and climbed the small bluff up to the bench to watch the trawler sail off. As the blue-hulled boat headed away, back to civilization, a bittersweet melancholy drifted over me. Our last concrete connection to the modern world was that boat, and in the churning of its prop wash, that connection was dissolving.

We were in a landscape of long, rolling outcrops, tundra plains and pockets, massive rock walls and glaciated peaks. The setting had the feel of a flooded Yosemite Valley: dramatic, austere, and beautiful. Small waves splashed along the cobbled shore, becoming a cadenced auditory backdrop.

Vaguely remembered serene experiences, filtered through years of longing to return, now confronted reality. The crystalline fjord water was bitterly cold; the rhythmic wash of water over stones made for lethally slick algal slimes; the beauty of the wild world was empty of affection. A lonely solitude blanketed the land as completely as the late-afternoon clouds covered the sky.

I walked from the bluff over to the rocky beach where we had piled our supplies and joined John and Kai, who were carrying food boxes, an emergency radio, tents, sleeping bags, backpacks, hammers, sample bags, and notebooks—the minimal necessities for our four-week expedition. John and Kai, in their inimitable way, had organized where each type of supply should be stacked, and how. In this wild place, some order was being imposed.

Kai was the anointed cook. His sturdy, rounded self spoke of a joyful respect for good food. He smiled often, and joked about how well we would eat as he strategically placed bags of onions and potatoes next to the cooking gear. Every food box was opened, its contents quickly evaluated, and a decision made about where it should be placed relative to the stove. Each of us enjoyed cooking, but for Kai it was integral to his spirit. Allowing him the privilege of cooking for us served us all well.

Much of our work would focus on rocks along the shore, where tidal scour had exposed, in cleanly washed surfaces, the patterns and minerals we had come to study. Such work required a Zodiac—an outboard-powered inflatable boat that could easily land on rocky beaches. John, the dedicated mechanic among us, unhesitatingly took on the role of Zodiac "captain." His beard, already growing into a grizzled stubble of gray and black, and his thinly lined face made him look the part. Taller than Kai and I, with a slightly gruff demeanor and dry humor, and a face vaguely reminiscent of the silent movie star John Gilbert, John projected an authority he embraced but did not demand. Invariably, he wore a blue baseball cap, which covered his very bald head, and a red anorak. Unlike Kai, whose strong Danish accent made it obvious where he was from, John had a deep voice that reflected the years he had lived in Canada, his accent attesting to a confusion of cultures. As I joined the two of them to organize our gear, John pointed to where each box should go.

Home was now a bench of tundra-covered rock a quarter of a mile long and two hundred feet wide, abutting a west-running ridge that disappeared under the ice. The late-afternoon Arctic sun was in descent toward the western horizon, struggling to warm the world cast in the shadows of a thick cloud cover and dusky light.

The permanent daylight was a liberation. Although the body's diurnal clocks are at first confused, and anxiety about whether or not sleep will be possible jangles nerves, an unexpected calm eventually settles in. The dictatorship of night's blackness, which constrains movement and limits sight, is banished. Clocks and time of day become unnecessary burdens. The freedom of timelessness seeps into life. We got used to taking strolls along beaches at two in the morning, with billowing globes of clouds backlit by the sun reflecting off of glassy fjord surfaces. Watching prowling Arctic foxes stealthily search for sustenance in the spongy tundra at midnight, clearly visible in the pale light, would become addictive.

AFTER UNPACKING, WE TOOK A BREAK FOR COFFEE. Kai put a pot of water over a hissing Primus set up on a flat stone. As we stood around waiting for the water to boil, red plastic mugs with a spoonful of instant Nescafé in our hands, we mused about our abrupt change in circumstances. Just twenty-four hours earlier, we had been in Copenhagen, one of the world's most sophisticated cities, where John met us at the airport for the flight to Greenland. Shortly before meeting John, I had been sipping cappuccino at a sidewalk café and enjoying the bustle of tourists along the quay in Nyhavn. I had flown in from San Francisco a few days before to help Kai finalize the logistics for the trip. Now, isolated from the rest of the world, removed from everything a "normal" day would bring, the meaning of *normal* became ambiguous. We were at the beginning of days of discovery, of seeing things never before seen. Excitement was implicit in every comment and laugh. The water finally boiled and Kai poured it into our cups, the smell of the instant coffee pungently punctuating the Arctic air.

But there was also an undercurrent of tension.

"It is nice to be back." Kai sighed as he looked across the fjord. His ruddy face glistened from the afternoon's efforts. A thin smile on John's face acknowledged the passing of decades. He was looking off in the same direction as Kai. I nodded, and uttered a slight "Hmm."

Across the fjord, nearly five miles away, a small ice field glowed white against the grayish greens and reddish browns of the tundra it rested on. We absentmindedly watched it as we mused about plans and what we might find. Eventually, Kai's comments turned to the controversy that had briefly been mentioned long before. He glanced down at the plant-covered ground and slowly ran a boot across it. He spoke with strong emotions about published interpretations of the geological history that conflicted with years of work and the field observations of two generations of researchers. He quickly alluded to the fact that those new conclusions were based on a single season in the field, and lacked the in-depth direct scrutiny earlier studies had benefited from. It was our task, he said, to break new ground, with more detailed attention given to specific locations and features that might resolve what was clearly a conflicted set of hypotheses.

I asked what papers he was talking about. Although I knew there was some disagreement about details of the geology—it is a science, after all, and debates keep things honest—there wasn't any specific paper I could recall that justified this attention.

John said that he had the papers with him and that he would bring them out later, his baritone voice taking on a serious tone. Then, breaking into a smile, he swept his hand across the scene in front of us. "I think this is a time to just be glad that we're back."

A few comments on the amazing beauty and feel of the place were made, but most of what was sensed was barely shared through small jokes and quiet nods. The emotions we felt were close to our hearts. After our break, we went back to work, setting up our individual tents.

By eleven, we were exhausted from thirty hours of travel and work. We said good night, headed to our tents, and climbed into sleeping bags.

Sleep came quickly, but I woke up within an hour. Tense from the excitement, sleep became impossible. I crawled out of my sleeping bag, pulled on clothes, an outer jacket, and boots, and slipped out of the tent. Shouldering the small backpack that was tucked under the tent's rain fly, I struck out for a hike up the ridge to our north to calm my mind. In the dusky light from a cloud-veiled midnight sun, colors and edges were muted, but the grandeur of the landscape was not diminished.

Arctic tundra, that unique organic collage of grasses, mosses, sedges, dwarf plants, and lichens, is often portrayed as dreary, as if it were a monotone of color and texture. But that is not the case. The tundra biome flourishes as a botanical riot, an evolutionary chaos rich with successes and possibilities—it is a deep velvet softening to the stone margins of a hard-edged world.

Mosses insinuate themselves into available spaces. Black, white, and orange lichens, their edges brittle and curled, cover in floret forms exposed rocks and branches. Arctic willows, ragged and squat, scatter about opportunistically, standing with quiet arrogance—at a height of two feet, they are the tallest plant. Flowers of white, pink, purple, red, and yellow are everywhere, sparkling like brightly colored gems scattered on a green-gray world. Clumps of cotton grasses, with their puffy white manes on waving eight-inch stalks, assert themselves with a graceful confidence.

Each plant extends roots into the decaying remnants of diverse ancestors, a living boreal flesh mantling thousands of generations of organic detritus. They huddle in hollows and drape over rocks, ponding water in small catchments, carpeting that cold world in a damp lushness.

Time is frozen in such a place. Whether I walked in a twenty-first-century landscape or a primordial, Ice Age epoch could not be told. That inability to know time riddles the experience of place, dislocating perception into an insecurity that, in my case, made it seem as though I had trespassed into some other world.

By the time I reached the first rock outcropping, the effort of pulling soggy boots out of the thick, wet tundra had grown tiring. My heart was pounding and I was breathing hard. Leaning against the twenty-foot stone face, I worked to catch my breath, rest, and expand my sense of what was around me.

The stone wall was nothing unusual, just the common gray layered and recrystallized gneissic rock we would see so much of over the next few weeks. Between the clusters of lichen colonies, bare rock lay open to the elements. I took out my hand lens and looked at a magnified rock face studded with broken crystals, excavated and sculpted by millennia of winter ice and summer rains. Perfectly shaped crystal faces and cleavage edges formed a microscopic, raw-edged sharpness to the rounded surface of the bedrock rib that was the ridge.

The scramble to the top of the wall involved a few minutes of light exertion, but it came at a cost. My fingertips, palms, and knuckles were bleeding in the time it took for that trivial ascent. I dropped my backpack, pulled out my gloves, and put them on over sore hands.

At the top of the small bench, I looked up and saw that the ridge I had seen from camp and thought was the top was only one of several shoulders below the actual ridge crest, which was several hundred feet above me. What was to have been a short saunter was going to be a longer hike. With a deep breath, I put the backpack back on and set off.

Walking through that land became a stroll along ponds of slowly seeping water deeply tinted with brown tannins, glistening. Some were enclosed in pillowed banks of deep green mosses, the somnambulant waters barely rippling as tiny streams trickled in and out. Other small catchments were little more than slight depressions in a saturated, vegetated surface. I could not escape the uneasy feeling that I was intruding into private gardens of invisible beings, constructed by them for the sole purpose of quiet meditation.

Moths and spiders and huge bumblebees appeared out of nowhere, gamboled about, and then instantly vanished. Flying creatures darted from flower to flower, briefly setting them in motion from the backwash of beating wings. But, except for the bumblebee, whose hum became a racket at close range, the visitors were silent.

Arctic wrens came and went, nervously concerned at my presence. They appeared out of the tundra from hidden places, fearfully attempting to distract my attention, worried I would ravage their nests. They had nothing to fear—I was incapable of finding those exquisitely hidden weavings of grasses and twigs.

As I walked on, up and over two more small shoulders and the intervening expanses of tundra, concern about the impact of my boots on that delicate place began to loom in my mind. Each step seemed an intrusion, punching down thousands of years of undisturbed growth in a brief, violently invasive moment. Guiltily, I turned to see the damage. It was stunning to realize there was nothing to see. With each step, that wet and soggy world yielded to the presence of a wandering mortal, momentarily exposing its most intimate details to daylight it had not known for centuries, but was hidden again as the boot was lifted and the yielding mass restored itself to its original form. In that world, I was no more significant than an afternoon breeze.

At first impression, the ability of life to thrive at that high latitude challenged reason and logic. But as the insignificance of my presence sank in, it became obvious that it was the toughness and tenacity of that living world that defined the reason and logic of the place. The biased patterns of thought I had inherited from another context were little more than low-level cosmic noise, a background hiss. I had yet to grasp the magnitude of my ignorance.

After perhaps thirty minutes, I reached the last wall of rock. Tired and sweating, breathing hard, legs burning, I climbed the last forty feet of outcrop.

The ridge crest was a slightly rounded, broad platform of nearly barren white and gray gneiss, randomly covered by the brittle lichens. I scrambled to the top and raised my eyes.

My breath caught in my throat. Extending from horizon to horizon, for nearly a hundred miles, untouched wildness rested silently in exquisite vulnerability. Stupefied, arms outstretched in submission, I slowly turned around, trying to take in the magnificence of the vista. Tears welled up as tangled emotions—sadness, joy, liberation, humility, anguish—flooded through me.

I turned toward the east and was surprised to see that the clouds ended at the land's edge, where it was subsumed by the ice sheet. Some mysterious atmospheric phenomenon demanded that, under the set of conditions that day contained, clouds that hung over land and sea would dissipate over the reflective frozen surface. Brilliant deep blue sky skimmed over the ice, framing the blinding white light of its crevassed surface.

From north to south, the sharp edge of the ice front zigged and zagged across the ground, marking a jagged boundary between worlds of conflicting expectations. In places, cliffs of white-blue ice soared hundreds of feet for miles, only to give way gradually to gentle ice hills and valleys that met the rock surface with slightly sloping indifference.

In contrast, the landscape to the north, west, and south was a mosaic of fjords, lakes, rivers, and mountains. The gray sky reflected off meandering waters, while the dark, shadowed land rose and fell in a pattern of parallel sharp-walled ridges. West-running fingers of ice-sculpted bedrock pointed toward the Davis Strait just over the far western horizon, the flow of landforms giving the scene a feeling of movement, a sense that some dynamic was playing out, even in the absence of motion.

To the south was the fjord on which we had just sailed.

That fjord, as with all fjords there, was cut into solid bedrock, confined to flow in narrow channels by sheer walls hundreds or thousands of feet high. Its breadth in places was more than five miles; in others, less than two. Although our camp was right at the water's edge, it lay hidden in the lee of that first small ridge I had scrambled up.

For long moments, I lived in a fantasy that no other person existed, that the lone human soul in all the world stood on that ridge, mesmerized by the bewildering wildness of everything surrounding him. As I stood there with those feelings, a vague unease settled in, one that would come and go throughout my time in Greenland. That feeling was not a sadness per se; rather, it was a quiet longing for things humanity has no words for, but with which wilderness settings overflow. There was a sense of missed opportunities, of an inability to connect with something profound, as though what I was immersed in shimmered incomprehensibly at the edge of sight.

OVER TEN THOUSAND YEARS AGO, during the last Ice Age, the landscape I stood on had been buried under thousands of feet of ice. Every valley and ridge that could be seen, every hillock and defile had been the floor to that ponderously migrating sea of frozen water. This was a young, inherited landscape, shaped by the grinding ice of that ancient time. As the Ice Age melted away and exposed the bedrock, the sculpted land provided footholds for pioneer plants. Season after season, as plant life slowly but

incessantly blossomed, withered, and died, plant remnants found anchor in ice-wedged cracks, lichen attached
to bare rock, and dust settled into pockets and irregulari-
ties, nurturing an unimagined future that included our
little camp.

As land plants took hold, Neanderthals and emergent
Cro-Magnons may have walked over the hillocks and
ridges there, searching for food and exploring. But it's
unlikely they settled anywhere in that harsh place—the
world farther south and across warmer seas was more
hospitable. Even so, it was difficult, when gazing at the ice
walls, not to imagine early humans skirting along them.

It was a panorama that defied comprehension. There
was nothing familiar there. The absence of trees, of houses
or streets, of cars or people, the lack of movement of any
kind—all contributed to a sense that I was walking alone
in an alien world, not of Earth, but of some planet where
forces and processes played out their dramas according to
different rules.

The longer I stood there, the more intense was the
conflict between the experience of the place and what I
had remembered of Greenland. As before, a deep sense
of serenity permeated everything that was present—there
was a unity of actions and substances, an uninterrupted
unfolding that shaped and colored everything. And yet,
something felt askew.

Then a lone bumblebee buzzed past my ear, soared
off into the valley, and disappeared, and it became clear
what that disjointed experience meant. Despite the

dynamism of that world, it was utterly and deeply still. I suddenly realized that it was the silence of the place that I had forgotten.

The gentlest breeze brushed my face, but there was nothing to hear. The distant rivers flowed, their shimmering surfaces vaguely vibrating with motion, but no sound emanated from them. I turned in every direction, listening for anything, but there was nothing.

What could be heard was the nature of the primordial world. Four billion years ago, on the barren surface of Earth's first land, with the exception of a rare roaring gale or exploding volcano, there would have been no sound. Similarly, in the ocean or the air, silence would have persisted, except where seas lapped onto continental margins and waves washed over eroding sands. In fact, for most of Earth's history, silence ruled.

With the emergence of animals more than 600 million years ago, that silent condition was slowly modified. Fishes clicked, bees hummed, dinosaurs roared and bellowed, birds chirped, horses whinnied, and, eventually, humans spoke and sang. The buzz on the surface that life brought to the world grew in complexity and volume, culminating in the constant roar of our cities.

A shout or a scream where I stood would have been swallowed in the expanse of wildness. That world was ancient beyond measure, holding on to the nearly vanished character of what once was, existing as a remnant enclave, speaking in its silence the song of our origins.

What was present in that vast, unimaginable panorama was an invitation to embrace anything and everything.

I stayed at the promontory for as long as I could, struggling to find a way to silence my mind. But my hands and feet were aching from the cold, and the exhaustion of the past day was beginning to take hold. Wrapped in the cloak of wilderness, I walked back to camp, trying to do nothing but listen.

THE NEXT MORNING, BEFORE I WALKED to the cook tent, I went down to the fjord to hear the sound of water lapping on the shore, seeking a connection to the world we had left behind. There was no wind; the surface of the water glistened like glass. The slight swell that slowly undulated that finger of sea did not stir a single grain of sand. What sound existed came from me.

I walked to the kitchen tent and joined Kai and John for coffee and our first breakfast for that field season. We explored the food crates, looking for appealing items, each grabbing his own unique mix of canned and smoked fishes, muesli, oats, powdered milk, bread, sugar, and jam. As we ate and planned the day, I kept to myself the walk I had taken. It wasn't the time to let them know I liked to wander off alone.

Mirage

W E WERE THERE TO OBSERVE, and to collect samples of anything that would provide evidence of the terrain's history—stretched crystals, folded and distorted rock layers, and any other indicators of tectonic movements. Noting on a map the places where each observation was made and where the samples were collected would allow us to piece together a tentative story while in the field. The samples we collected would be shipped back to our laboratories, where we could later assemble other facets of the history—how hot the rocks had been and how deeply buried they were when the deformation had occurred. The field observations combined with the results from the laboratory would provide the factual framework for the history we would write of what had happened thousands of millions of years ago.

The vanished mountains we envisioned were simple possibilities, tentative interpretations of passages written subtly in the obtuse patterns and features of Greenland's rocks. The patterns match those seen in the Alps and the Himalayas—zones that seemed to be huge thrust faults, folds of immense proportion, metamorphism at extreme conditions. Through the inspired power of analogy, Kai,

John, their coworkers, and those who had come before them had surmised that the Greenland landscape was an old ancestor, a forerunner of the young mountain systems that today so dramatically exalt Earth's skin. But the Greenland ancestors are long gone, erased by the incessant hunger of flowing water, blowing wind, and grinding ice to achieve a form of topographic equality between sea and land. Erosion always wins.

The first clear hint of those lost mountains had come years earlier. Just after World War II, the Geological Survey of Greenland (GGU) was founded in Denmark. Through its offices, a small group of geologists, including Arne Noe-Nygaard and Hans Ramberg, began the first systematic study of the west coast of Greenland, sailing along the complex coastline in motorized sailing vessels strengthened to resist collisions with ice. They found a two-hundred-mile-wide belt of rock that seemed to preserve evidence of multiple complex episodes of protracted and intense deformation. The belt was called the Nagssugtoqidian mobile belt, named for the region it cut through—Nagssugtoq—and the fact that the rocks seemed to have been twisted into structures that implied extreme plasticity and flow. The mobile belt ran east-west all the way across Greenland. Although the mobile belt seemed to represent a major orogenic, or "mountain-building," event, how or why it formed remained enigmatic. Cutting through this region were several distinct zones, each zone a few miles to tens of miles wide, in which the rocks were steeply inclined and consistently aligned in the same direction. For some years, the significance of the zones of aligned rocks remained obscure, their tectonic significance unknown. But by the late 1960s and early 1970s it had been suggested by Arthur Escher and Juan Watterson, among others, that these zones contained rocks that had been severely sheared into steeply inclined parallel sheets and layers. The individual zones were eventually called shear zones and were named after the regions they ran through—Isortoq, Ikertoq, Itivdleq, and Nordre Strømfjord. The latter, the Nordre Strømfjord shear zone (NSSZ), became the center of attention because it marked the northern edge of the entire Nagssugtoqidian mobile belt. It was the only one for which observations were made near the ice—the others were only mapped while sailing along the coast, and their inland extent was unknown.

Geology is not generally considered an enterprise rich with drama. Rocks stolidly await inspection, slowly providing, through insightful consideration, a glacially paced story of incremental change. But there are occasions when perspectives are radically altered, new story lines emerge, and the field is caught by surprise.

In 1987, such a change shook the world of Greenland geology. Although it played out subtly, the consequences for all involved were profound. Feiko Kalsbeek, Bob Pidgeon, and Paul Taylor reported finding along the northern limits of the mobile belt, near the inland ice, remnants of the same type of rocks as those found today in the Andes

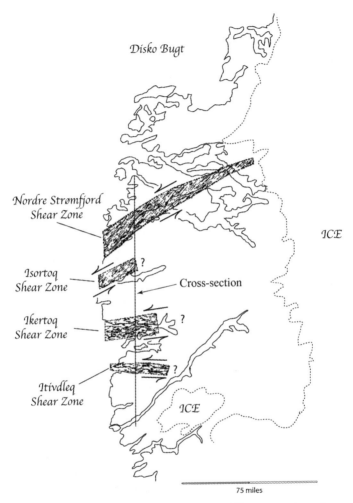

The shear zones of Escher and Watterson. The arrows show the inferred direction of movement on either side of the zones. The vertical line shows the location of the cross section in the figure on page 60. Modified from a drawing by Kai Sørensen.

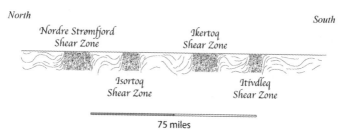

North   *South*

Nordre Strømfjord   Ikertoq
   Shear Zone   Shear Zone

        Isortoq   Itivdleq
     Shear Zone   Shear Zone

                    75 miles

A cross section, ca. 1976, showing how the shear zones disrupt the otherwise gently deformed layering in the rocks in the figure on page 59.

and the Sierra Nevada range in California.[*] Although nearly 2,000 million years older, those rocks were evidence that what is happening in the Andes today had happened in Greenland. In the case of the Andes, the continent of South America moves west, riding over the floor of the Pacific Ocean and pushing it hundreds of miles below the surface. Plunging into the incandescent heat of Earth's interior, generating massively destructive earthquakes, the ocean floor partially melts, giving rise to bodies of molten rock that slowly make their way back to the surface. The volcanoes of the Andes and the mountainous spine they decorate are the result of that process. If the analogy was accurate, somewhere hidden within the Nagssugtoqidian mobile belt there should be evidence of a vanished Pacific, but no evidence of such a thing had yet been found.

---

[*] F. Kalsbeek, R.T. Pidgeon, and P.N. Taylor. 1987. Nagssugtoqidian mobile belt of West Greenland: a cryptic 1850 Ma suture between two Archaean continets—chemical and isotopic evidence. *Earth and Planetary Science Letters* 85:365–385.

Kalsbeek and his coworkers acknowledged the enigma, and suggested the ocean may have been swallowed in the collision of two small continents. Such a concept had the power to explain the significance of the mobile belt and the major fault zones in it—the structures reflected the massive deformation expected as a result of two continents colliding head-on. But the evidence for where the actual collision zone might have been was very sparse—there was no good way to identify where the rocks from the old southern continent ended and the rocks of the northern continent began. Compounding the uncertainty was the underlying debate of whether plate tectonics even functioned that long ago.

The areas where John and Kai and their colleagues had worked were central to answering those questions. The evidence they had developed suggested that the collision zone, which would have required exactly the same kind of massive movement and deformation they described, might be within the areas they had worked.

Those who study the history of Earth are few, and the areas involved are vast. Knowledge is sparse. Given the immensity of the terrains the continents cover, those dedicated to unraveling the story of evolving landscapes devote their lives to finding the nuance and subtlety held within a specific setting. Some spend their lives immersed in the history of the Alpine system, climbing and hiking through those beautiful mountains. Some are owned by the Himalayas, or by the vast openness of the Canadian shield. For John, Kai, and me, it is Greenland.

Inevitably, commitment to place becomes personal—our identity is affected by the time we spend walking the fragment of Earth that has captivated us. The chosen place permeates being—terrain embeds itself under fingernails, tangles in hair, makes skin bleed and scars the heart and mind. Every thought, conscious and not, becomes riddled by knowledge derived through wandering there; remembered vistas from that world unexpectedly insinuate themselves at random times and in unanticipated ways, forcing an acceptance of a link between what we experienced there and what is lived in the moment here. We are composed of where we have been and what we have seen.

John and Kai were part of a pioneering generation that helped refine Greenland's history. They and their colleagues described in detail the characteristics that defined the "mobile" part of that land—the folds and sheared layers, the discontinuities and disrupted features. Over the years, they mapped major tectonic elements, documenting evidence for miles of displacement along several of the shear zones. They published respected papers in scientific journals, and were recognized authorities because of their work. They knew that land better than anyone. But in the late 1990s their reputation as field geologists and scientists was challenged by a paper that said, in essence, the work they had done was deeply flawed.

THE GEOLOGICAL SURVEY HAD FIELDED a number of small teams in a very large area and it had been decided that each team would check in by radio with a base station at Aasiaat each morning and evening. That would allow a rapid helicopter dispatch if an emergency happened. It was the first and last year in which we checked-in by radio, since, in later years, we were the only team in the region and the cost of a base station could not be justified. On the second evening of our expedition, we were confronted with an identity crisis. For our first radio check-in, we needed a name to identify ourselves so we would not be confused with others in the field. We were the last team to be placed in the field that year and would be among the last to leave. Before our arrival, each team had taken a name to call in with, and we needed something unique.

We quickly tried to come up with a snazzy name—call-in time was fast approaching. When the moment arrived, John and I looked at Kai and shrugged our shoulders. Kai pursed his lips, clicked the button on the microphone, hesitated, then said, "This is Team Alpha to base, over."

There was a pause at the other end; then base responded, "Come in, Team Alpha. Welcome!"

After the call-in, we asked Kai why he named us Team Alpha. He said it came to him that we were the oldest folks in the field, making us the alpha males.

LATER, AS WE WERE SITTING IN THE COOK TENT talking about plans and challenges while Kai prepared a chicken—the last fresh meat we would have for weeks—I asked about the comments from the previous evening that seemed so emotionally laden. Quickly, the mood became more serious. John looked at Kai and Kai nodded. John reached into his stash of reading material and handed me a seventeen-page paper published five years before.

The paper asserted that Kai and John, among others, had made basic and fundamental mistakes reading the rocks. The new publication stated that the NSSZ showed very little evidence of significant movement. It said that in a collective misinterpretation an essentially trivial feature had mistakenly been given major tectonic significance. The words *shear zone* were removed from maps in the paper and replaced with *straight belt.*

Science is a messy business; everything we know is, at best, a simplification of what is real and is therefore inherently flawed. As a consequence, everything we do ultimately requires corrections, implying that nothing published is completely right. It is every scientist's expectation that whatever he or she publishes will be improved upon by others, who will provide more nuanced and detailed observations that address questions about the world. Indeed, it is an honor to be a building block in an ongoing refinement of the story of how a landscape has evolved. But in the case of the paper I was reading, it was difficult to escape the fact that Kai and John's work had been summarily dismissed.

About halfway through reading the paper, I stopped to ask them if they agreed with what it was saying, that they had been wrong about how they had interpreted the geology. "Of course not!" was the answer. At first, they spoke with disciplined calm. But quickly, with increasing emotion, they signaled numerous inconsistencies and errors in the paper, fundamental mistakes and misinterpretations that exceeded what the paper itself had, inaccurately, called to task. But only those intimately familiar with the real rocks would ever know.

Kai pointed to a black-and-white photo that showed horizontal layers on the face of a cliff wall. The text of the paper interpreted the geology as flat-lying layers in gneiss, indicating structure that was incompatible with the model of a shear zone with nearly vertically inclined fabric. "You've been there, Bill. Do you remember? Those aren't flat-lying layers!"

At first, I did not recall the location or the rocks. Kai then commented that I had seen it during my first expedition to Greenland, at a site where the edge of the shear zone was being investigated. Memories then flooded back.

We had camped on a small finger of water on the south shore of Nordre Strømfjord. To introduce us to the geology we would be working on, Kai took us on a daylong hike away from the fjord and into the country that bounded the southern edge of the shear zone. All the rocks where we were camped were vertically inclined gneisses with bands of dark and light layering that varied in thickness from inches to many feet. All the layers trended in an east-northeast direction. The hike he took us on was across the layering, heading south. Since there was no trail, the route he chose followed streams and small valleys. The cliff wall shown in the photo in the report was bounded on its west end by one of the valleys we had followed. As we walked by the end of that ridge, we could see that exposed on its barren surface were steeply inclined, but not vertical, dark and light bands. Kai stopped us and pointed out that the farther south we would go, the less steeply inclined the layering would be. Where we were was the southern edge of the shear zone, the place where the dark and light layers had been progressively rotated and twisted into parallel orientation with the main fabric in the central part of the tectonic belt. The reason the layering looked horizontal in the cliff wall depicted in the photograph was because the cliff face ran exactly parallel to the trend of the layers, not across the layering, where the steeply inclined orientation could be seen.

One learns in any introductory geology class that what one sees in the field requires careful observation and measurement in order for one to understand what is actually there. The land we walk over is a three-dimensional surface that intersects complex geological structures. Piecing together the actual form the geological bodies possess requires following them across ridges and valleys, mapping them, touching them, and astutely observing how the land surface and rock forms affect what is seen. Clearly, the published photograph had been taken at a distance from some shoreline vantage point or cruising boat and thus was not part of a field excursion to verify an interpretation.

Consequently, as things stood in the international scientific world, the work Kai and John had published was implicitly worthless and could be seen as one more example, among thousands, of failed scientific ideas.

When I had finished reading the paper and began discussing with Kai and John the scientific conundrum we were in, I realized the devastation and angst they must have felt. I had known these men well for many years; I had watched them argue and debate, examine data and analyses, discuss conflicting ideas. I knew them to be critically thoughtful. John was a data-driven man, always examining information through the lens of logic and rigor. He was not a sloppy scientist. Kai was a grand thinker. He had worked long and hard at piecing together fragments of information into concepts and models that could explain mountain systems. He had studied the giants of geology, those who had made huge leaps in thinking about how Earth evolved. He could see patterns and relationships that were often, at first, vague and equivocal. But his ability to weave together the threads that unified a fabric was brilliant. To think either of these men could have been so misled in formulating their concepts was inconsistent with everything I knew about them.

Being the rigorous scientists they were, they framed the argument for our little expedition as a data-gathering effort to resolve the conflict. At the time they invited me, they had said the purpose of the expedition was to pursue unanswered questions. There was no doubt that was, in fact, the underlying justification for the work. But I also realized this was, in part, an expedition for their own vindication.

THE MORNING WAS VERY STILL, befitting the first day of work after the previous night's soul-rending honesty. Even though the sun blazed in a deep blue sky, the air temperature was close to freezing. Kai and I sat in the bow, huddled against the wind as the Zodiac sped down Arfersiorfik Fjord. I pulled the hood of my anorak over my head and put on gloves. Water splayed off to the sides in fragments of refracted sunlight, decorating the mirrorlike water surface. The outboard roared. John had the throttle wide open.

We were headed for the northern boundary of the Nordre Strømfjord shear zone, which had been approximately mapped many years ago. Very little detailed work had been done there, mainly because it was so remote and difficult to get to. On our maps, the edge of the zone was confidently drawn in black ink, but we knew that no one had actually been there.

We sought that tectonic landmark as a reference point, a location where the fabric and grain of the rock could be seen and felt. We were searching for something that could be quantified and analyzed, something that would establish metrics for later measurements and comparisons. In order to be able to recognize severely, as opposed to minimally, sheared rock, we needed a baseline.

The three of us gazed down the fjord as we flew across the translucent water. Despite the roar of the outboard, we were enthralled by the beauty of the place—the hills rolling gently to the sea, the flower-chocked rivulets cascading down the bedrock, the stillness of the scenery. With some effort, we tried to focus our attention on the rock wall to our south, with its extensive exposure of folded and sheared gneisses.

Unexpectedly, as we watched the steep walls of the southern fjord edge, something shifted far to the west, down the fjord and miles away. I turned my head to get a better look, but at first all I felt was confusion. Initially, I thought the distorted landscape I was seeing was due to my eyes watering in the cold wind, but after rubbing them I realized something extraordinary was dancing along the horizon.

The land on the north side of the fjord was broad and rolling. Soft ridges sloped down to the water in a subtle cascade of rocky knolls and tundra pockets. It was a landscape that invited daydreaming. In the early-morning sun, the scene looked almost pastoral.

But farther down the fjord, a thick horizontal blade of sharp turquoise blue cut across the land, as though a giant painter had saturated a brush and slashed the ground with it. The blue was brilliant and intense, a pure distillation of color. It seemed to stretch hundreds of feet into the air and was painted across the land for miles. Within that absolutely horizontal turquoise stripe floated vertical columns of white, gray, tan, and green, looking for all the world like skyscrapers in a city miles away—a shimmering blue Oz resting on the frigid waters of the fjord. Toward the north and east, the blue trailed out into a needle-thin line that vanished at a piercing point sharper than the edge of a razor blade, ending in the middle of the rolling hills.

We all saw it. As we cruised, we watched immense rock masses from the rolling land split off and drift into the blue blade, becoming the skyscrapers that floated in the air. The size of the masses was staggering, seemingly miles wide and hundreds of feet high. As they drifted slowly out into the fjord, they changed form, shifting from angular columns to smoothed elongations filled with textures and patterns, never resolving into a constant shape, and then slowly vanished—evaporating as though consisting of nothing more than mist. Eventually, the effect was too stupefying. John throttled down the motor, the bow dropped, the roar of the engine stopped, and we drifted with the tide.

We sat silently for minutes, watching the fata morgana while the Zodiac slowly turned and drifted in the gentle current.

A nearby island only a few hundred yards away subtly entered the scene. The knob was a small rocky knoll, covered with mosses, shrubs, and lichen. On our maps, it was an ink dot so tiny, it wouldn't be noticed unless one were looking for it. As our line of sight shifted to the point where the small island came between us and the mirage, regret began to well up at the thought of losing that magnificent show.

Without preamble, and with extraordinary understatement, the distant blue line slowly sliced across the small island. The effect played out with such surgical precision that the inconsistencies between expectations and experience took a moment to register. Emphatically, right in front of us, the little island was divided into an upper and lower half, sandwiching a thin brilliant turquoise layer.

I struggled to accept what my eyes were seeing. The implication was obvious and inescapable: What had seemed so immense and distant, miles down the fjord, was little more than a pencil-thin, trivial mirage barely an arm's reach away, hovering in the air like a butterfly before my nose, somewhere between our little rubber boat and the small rocky knob of an island.

In that moment, what we knew to be true because we had seen it in the company of others, suddenly became unequivocally false, for all of us. But we did not have the luxury of time to resolve the contradiction. A distant destination waited, offering an opportunity to collect desperately needed data, and the afternoon winds would surely come up, making it difficult to get back to camp. Without discussion, John started the outboard and we continued on.

As our vantage point changed and we rounded the little island, the mirage returned, immense, awe-inspiring, silent. It stayed with us for ten minutes more, then slowly melted away into the thin air.

COLD DENSE AIR, CHILLED BY the frigid fjord water, had refracted light, bending it into a vision. Light is a malleable thing, warped and distorted by well-known effects, conditioned by a broad range of circumstances. What we are able to sense, which is less than one billionth of a billionth of the electromagnetic spectrum, is affected by the sensitivities of the organs our bodies use to detect it, and the narrow range of physical conditions within which we wander. Despite the richness and beauty of the things we can perceive, we remain profoundly impoverished by the limitations of our genetically constrained bodies and the space through which they move. What we see of the world is our own manufactured carnival—the mysterious unknown within which that carnival resides beckons through mirages, silences, and misunderstood truths, forever beyond our grasp.

What I did not realize at the time was that mirages are visual earthquakes, sometimes of great magnitude. Such ruptures arrive with a low rumble a few moments before the ground shaking starts. If one is aware, attuned to such powers and their potential for wreaking havoc, the direction the rumble comes from can be perceived and a quick adjustment may be possible in order to brace for the impact. But I was not aware and did not recognize the implications. What was to follow over the weeks and months in that wild place was the ground shaking of self.

We did find the northern edge of the zone that day, but it was not where we had expected it to be. The black-inked line that marked the edge on our preliminary maps was off by miles. We also discovered types of rocks we had not expected. What all this meant was not at all apparent; it led to many arguments that resolved nothing.

It was, too, a subtle warning. Lines on maps suggest boundaries, and boundaries shape expectations and provide limits; they simplify and categorize, making it easier to react without thought. The natural world, though, is flow and process, not limits. What we place on a map is an approximation, at best, a way of saying that things here differ from things over there. If we were to truly understand the place we wandered through, sampling and measuring and recording, we needed to respect the implication that boundaries are simply another form of illusion.

Breaking Stones

THE QUESTION OF WHAT HAD HAPPENED within the Nordre Strømfjord shear zone nearly two billion years ago danced through every waking moment. Was there a place, somewhere along the ground we walked, that was the first point of contact where continents had collided? What would be the sign? Or was the vision of entangled landmasses a flawed story, a misinterpretation of history? Regardless, how did the shear zones or straight belts fit either tale? The trip to the northern edge of the shear zone had added more observations and hard data but lacked sufficient context to inspire imagining.

For relief from the wondering, we would occasionally take a short stroll together around the hillocks and along the beaches near camp. These were casual and slow hikes, a chance to talk and look at things in an unhurried way. Anything we found could easily be revisited, so we took with us only hammers and hand lenses and notebooks, the minimum equipment necessary to descend below the surface if that seemed necessary.

One particular day, not long after setting up camp, we headed west along the shoreline in the late afternoon. There was a mile of land we had not seen, and we thought this would be a good way to familiarize ourselves with details and patterns.

Almost immediately, John discovered a spectacular example of what we came to call "pencil gneiss." The rock was the same type of igneous rock that had inspired Kalsbeek and his coworkers to propose the idea of a collision zone, or "suture," between continents, but there, where John stood, the delicate textures that form in slowly cooling magmatic bodies had been smeared into pencil-like forms, stretched and elongated. Individual crystals that normally were equant and half an inch in size had been strung out like taut pieces of string into thin lines several feet long, each precisely parallel to all those around it—a metaphorical pencil in the gneiss. That was graphic proof of extreme shearing. We took pictures, made notes, and placed another imagined factual stake in the ground. The immediate question now became whether or not such features were throughout the shear zone, or simply local and thus not of regional significance. We walked on, amazed, wondering what would be around the next headland.

A few hundred yards farther along the shore, we came upon a bizarre little cliff face. Hazy, dark lines patterned the surface, looking much like a pile of slightly deflated and sagging soccer balls stacked one upon another. We pored over every inch of the outcrop, struggling to piece together a picture of what we could not quite make sense of. We debated options and argued, running through every idea we could dredge from our experiences. What repeatedly came to mind was a jumble of tears, caught in the instant they were shed, as though Earth had wept from some unseen eye.

Grudgingly, we agreed the most likely answer was that we were looking at a deformed slice, perhaps 150 feet long and 50 feet wide, of a type of volcanic rock called pillow basalt, which forms when lava erupts under the oceans. Unlike the rocks surrounding them, which preserved evidence of complex histories with multiple episodes of folding and shearing, the pillow basalts had a very simple history: They had erupted onto the floor of some ancient sea and then been metamorphosed and simply folded once. That slice of rock was a lens encased in the much more intensely deformed shear zone gneisses and schists. The contrast with the surrounding rocks was dramatically obvious.

If that interpretation were true, the implications were staggering. Ocean basins the size of the Mediterranean or Atlantic commonly separate continents. If the continents are approaching each other, the ocean floor between them is consumed along the boundary that will eventually become the collision zone when the continents run into each other. Such collisions grind on for tens of millions of years, slowly exuding sheared, twisted, and recrystallized rock that had once been the sediments and volcanic pillow basalts of the seabed. It is from such "root" zones that Alpine-like mountain systems emerge. If that folded pile of pillow basalts we had just found was, indeed, all that was left of some long-vanished ocean basin, we had found the suture. That thin remnant slice was all that remained of what once had been a sea probably thousands of miles wide. Could it possibly be that we had stumbled upon the long-sought ocean that, fifteen years earlier, Kalsbeek and his colleagues had postulated might have existed there?

The excitement over that discovery was tempered by a healthy skepticism. Each of us had the experience of interpreting a fact or observation as evidence for some grandiose concept, only to see it crushed under the weight of more data and observations. We held little confidence that one outcrop would be the cornerstone piece of evidence supporting the ocean-floor idea, but neither did we dismiss it as meaningless.

Several days later, along the same trend and a mile west, we came upon another small slice of rock that showed exactly the same simple history preserved in the pillow basalts. It was a different rock type, though, called peridotite. Peridotites are the source rocks from which basaltic lavas are generated, and the rock type we were seeing was precisely what geologists associate with lavas erupted on the ocean floor.

Although it was seeming to be more likely that we had stumbled upon the true collision zone, two outcroppings of rocks are insufficient evidence to allow much certainty for such an imaginative leap. The history of a mountain system is a long story, told in many chapters. An outcrop is, at best, a paragraph in a chapter. We were historians, trying to read ancient texts written in a language we barely knew. But something was being revealed that had not been seen before. There had been tremendous deformation and movement within this zone, part of it involving the consumption of an entire ocean basin. And yet no one had suspected the existence of the collision zone there. It now seemed, between the pencil gneisses and these two new outcrops, that John and Kai would be vindicated.

The satisfaction Kai and John felt was obvious but muted. They remained thoughtful in how they analyzed everything we observed, but the edge was off. We found many more examples of the pencil gneisses along the trace of where the shear zone should be, providing irrefutable proof that intense deformation was distributed all along it. But the two slices of what might be ocean floor within the same belt of rocks made the story much more complex. We had not anticipated the finds suggesting ocean crust might be actually within the shear zone, implying the shear zone itself was the suture.

IF THOSE DISCOVERIES OF PUTATIVE ocean-floor basalts had tectonic significance, there had to be something more to that story hidden in other sites where contemporaneous rocks were outcropping. So we moved our tents to a location miles west of the pillow basalts, establishing a base camp along the same trend of rocks in Ataneq Fjord in a place for which no data existed.

On a day when a soft breeze blew over the sea and the sky was radiant, we headed out in the Zodiac to a place near the head of the fjord, miles east of the campsite. There was a freshness in the air that inspired a sense of lighthearted expectation that we would find something significant. We cruised smoothly over the clear water, watching the serene tundra-laced ridges and valleys and the low, rolling hillocks glide past.

We sailed for some miles, then landed on the north shore to walk the exposed outcrops. The tide was ebbing, revealing a pebbled sandy beach. John swung the bow around, cut the motor, and we slid up on the sand. I jumped out and tied the line to some boulders. We grabbed our rock hammers and backpacks and began walking east. Soon, I was distracted by some unusual rocks riddled by small veins of once-molten magma. I stopped to look and sample. Neither Kai nor John had much interest in what I was pursuing, so I told them to go on ahead and I would catch up.

I stayed there for perhaps ten minutes, then continued along the shore, enjoying the solitary stroll in the late-morning sun. Little waves lapped at the shore to my left. A light breeze blew, making mosquito netting unnecessary. It was almost warm enough to take off my anorak, but not quite.

After a short hike, I came upon a glittering bluff that stood like a white wall against the edge of the pebbled beach. The rocky surface was covered with fine, thin threads of white sillimanite crystals, barely discernible to the naked eye, all aligned in a flowing arrangement of undulating near-parallel fibers. Densely scattered within that white fabric were deep red garnets the size of golf balls. Pale mica and black graphite flakes glittered in the sunlight, disseminated within that wavelike surface, giving the outcrop the feel of a moving, rippled skin. For a moment, I felt as though I were in an art museum, gazing at a masterpiece that had been conceived and executed by some transcendent soul dedicated to making things of beauty. I walked over to the wall and reverently ran my hands over it, the garnet lumps bumping against my fingertips—all the while feeling as though my touch were a desecration.

A sense of irony then slowly took shape around the garnet clusters, the shimmering white crystalline threads, and my invasive fingers. Where I stood, an extraordinarily beautiful collection of shapes, forms, and crystals rested in the warming sun, exposed for view in the middle of a wilderness landscape so vast that there was little chance they would ever again be touched and seen. And yet, the seeming reality of the place was that the glittering wall was nothing more than the mundane solidity of a bedrock outcrop. How odd it seemed that only the thoughts conceived in a feeble brain, guiding the movement of dirty fingers, made that bare stone wall so beautiful.

The minerals sparkled in the sunlight, their shimmering, stunning patterns irrelevant to the lapping waves and gentle breeze. I took my camera from my backpack to photograph it but then decided against it and put it away. What would be the point? The reality worth capturing was the feeling of the place, the tender passion of awe-thrilled connection to an exquisite rock wall forged in the deep earth aeons ago. It then occurred to me that all was equal here: An absence of hierarchy reigned, everything was beautiful and not. Value relies on scarcity and a desire for difference, neither of which had meaning here.

Strolling along the gravelly beach, the only sound that of small splashing waves and the crunch of my boots, I was where I craved to be, walking alone in the presence of wild solitude, the treasure of aloneness seeping from the sunlight, from the blue waters, from the patterned stones. I have loved such places for as long as I can remember. When I was young, solitary walks in the hills near home provided an escape from bullying and rejection, a small child's refuge, where disappointments could be hidden behind the smell of sun-warmed grasses, the buzz of insects, a sudden glimpse of a vanishing snake slithering through the weeds. The experience of discovering hidden things, a ladybug behind a curled leaf, a sand crab dug up from an empty beach, fertilized my imagination. And now so did the white rock wall.

A SHORT WHILE LATER, I caught up with Kai and John. Kai, the primary recorder of our observations, was writing in his notebook, occasionally placing the graphite tip of a short pencil stub on his tongue. In his left shirt pocket nestled several other pencil stubs, his favored writing and drawing instruments. Where he gets the stubs, I have never figured out, but he is never without them. A small sharpener is always in another pocket.

I asked excitedly if they had seen the small bluff of

garnet-sillimanite schist, to which Kai responded per-functorily. He showed me the brief note he had made in his notebook about it.

He then asked, "Did you see that lens of greenish rock that is probably ultramafic a few hundred meters before that?"

I struggled to recall seeing it but had to admit I hadn't.

"Surely you are pulling my nose. You should go see it. John thinks it's significant." Chiding me was a revered pastime.

"It's pulling your *leg*," I said, correcting him. Kai's fondness for idioms was well known, but his use of them was occasionally flawed.

As I turned to go, John, who is an exceptional field geologist, commented that it looked like a tectonic slice of peridotite.

I had no problem finding it; the pod was exposed on a bare bench of rock that made up a small promontory in the fjord. The yellowish green mass was small, perhaps six feet wide and twenty feet long, surrounded by alternating light and dark layers, and obvious.

The pod was indeed a peridotitic body. Peridotitic rocks do not normally occur with sediments, which was what the garnet-rich rocks had originally been. Their juxtaposition against one another required mixing by intense tectonic forces. Here was even more evidence supporting the "vanished ocean" hypothesis.

Crawling over the outcrop, looking in detail at textures and minerals, one layer in particular stood out. It was three feet from the ultramafic body, about six inches thick, nearly black, and perfectly paralleled the edge of the yellowish green mass. Although it appeared that there might be garnets in it, they were tiny and impossible to actually make out. We needed to collect a sample.

We each carried two rock hammers—one that weighed a few pounds and was used on most rocks, and the other a sledgehammer that weighed about five pounds and was used for rocks that were particularly tough. The black layer stood a couple of inches above the surface. It was obviously resistant to erosion, and looked to be particularly dense, so I grabbed the sledge.

I have pounded on rocks around the world, but that was, without a doubt, one of the hardest rocks I have ever encountered. With every blow, the steel of the sledgehammer rang loudly and bounced off the rock. I swung harder and harder, fearing that the thick wooden handle would splinter at any moment. Eventually, a hairline crack appeared and began to grow larger with each blow. With sore, stinging hands, I finally was able to wedge free a small sample barely the size of my fist.

That small sample was unusually dense. The fresh, new surface glistened like broken glass, fine-grained and compact. I got out my hand lens and brought the sample close to my face so I could examine its mineralogy in detail. Suddenly, a faint smell like that of singed hair, hot metal, and desert dust wafted into the air from the fresh surface. Startled, I stopped what I was doing and breathed deeply.

There was no doubt: The smells were there in the air, rising from the surface of that newly exposed, sparkling face.

Hammering on that rock had broken chemical bonds that held it to the face of the outcrop. Tiny crystals had cracked, grain boundaries separated, and a very dense rock had fractured. For the first time in over two billion years, the atoms and molecules trapped in that crystalline framework were exposed to fresh air and the warming rays of an Arctic sun.

Displaced and broken, submicron-size particles and inorganic molecules had flown from the fracture, dancing in the air in an unseen atomic ballet, moving to the vagaries of a gentle breeze. A small fraction of those liberated pieces wafted through the atmosphere, traveled toward my face, eventually affecting the sense organs in my airways, stimulating unexpected and incongruous sensations—the impression of singed hair from a broken fragment of rock? Hot metal? Desert dust?

That fractured surface had spilled carbon and calcium and magnesium atoms into the world in a violent act motivated by curiosity. Everything that made that rock, and which would normally be released to the oceans through excruciatingly slow erosion, had suddenly been thrown to the wind. The atoms of that layer were the components of molecules that make life possible—everything from sodium to selenium had exploded into the breeze. Thoughts and imagination flow in tangled networks of neurons and synapses fed by the chemistry all those elements embraced. The potential for dreams was there in the atoms of that rock I was smelling.

What form the atoms and molecules would eventually take was an unknowable mystery—and little more than one more part of a long and timeless journey. Inevitably, once released, they would become part of something new, something wholly different from the mineral structures of which they had just been a part. The destructive act of collecting that small sample was, in a minuscule way, an act of liberation and creation, an unintentional and naïve perturbation of the future.

I picked up the sample and labeled it "468 416," and took a few photos. I pulled out my GPS, recorded the location in my notebook along with a few observations, and then stuffed everything in my backpack, not suspecting that little sample, once analyzed in the lab, would shatter our preconceived concepts about the history recorded in those very ancient rocks.

Cladonia Rangiferina

Lichen abound in Greenland. Above the tidal zone, every bare rock surface is textured and colored by lichen clusters, splotches, and mats. Tundra pockets will have lichen threading throughout them. It is the hardiest of partnerships, a pervasive symbiont of fungus and its photosynthetic companion, living as a composite organism, as resilient as it is beautiful.

There are many different forms, but my eye, trained for minerals and rocks, and not the things that grow on them, discerned only a few. Pale green, brilliant orange, and reddish brown varieties are there, intermingling in fantastic patterns of free-form, organic composition, a background pattern subtly embossed on hard rock. They carpet and upholster, embellish and decorate in a profusion that beguiles the senses. They draw you into hidden worlds where, facedown and wide-eyed, dramas can be invented, played out by tiny bugs wandering through lichen-framed halls.

Lichen are, as well, a hazard for the careless. One particular lichen introduces itself assertively. When it is dry, its deep black, frilled platelets are extremely brittle—if stepped on, the fringes along its margin crush with a crackle to a fine powder. If touched with a naked hand, its edges cut. But when it is wet, it is like mucus. On drizzly days, it soaks up water, becoming a mat that is impossible to walk on without slipping and falling. Once, when we were landing at a barren rock outcropping, I took the bowline in hand and was getting ready to jump from the Zodiac onto shore. John shouted over the noise of the outboard, "Watch out for the litchen [the Danish pronunciation he and Kai used for the word]. They are slippery!"

Acknowledging his warning, I adjusted my plan, picked out the flattest spot with the least amount of the mucuslike masses, and leaped very carefully, trying to land with as little forward momentum as possible, but my feet slipped the instant they touched the slime and I landed hard, briefly dislocating my right shoulder. I was on massive doses of aspirin for the next three days.

THE LICHEN ARE ALSO MARKERS OF SORTS. They grow slowly, even under the best of conditions. A growth rate of a thirty-second of an inch a year is fast. In the Arctic setting we were in, they grew much more slowly.

One dry, sunny day, we stopped on the south shore of the fjord, at a place where a gently sloping gneiss outcrop came down to the water. We were looking for the contact between two different types of rocks we had found the day before farther up the fjord. Above the high-tide line, lichen grew in profusion, particularly the black variety. As we walked along taking notes, we came upon a place where people had scraped away the lichen, leaving their names and dates in a negative space. All the dates were pre-1960; the oldest was 1943. The names and numbers were clearly legible, the edges barely changed since those messages had been scraped into the surface decades earlier. Those lichen were growing at a rate that could not have been much more than 1/1000th of an inch a year.

One lichen that did grow more rapidly than that was *Cladonia rangiferina*. That lichen is cream-colored and tends to form small, fringed, branching shapes that slightly stand above the tundra background. I first saw them when we set up our camp. I asked John what they were—he was a remarkable store of information about the landscape (some of which, I suspect, was made up, although most wasn't). He had taught me how to recognize old campsites—stone patterns and concentrations of certain grasses that favored disturbed ground—among other things. John said they were called reindeer lichen because the barren-ground caribou that wander throughout West Greenland, and on one occasion actually walked through our camp early one morning, eat them as an important part of their diet.

SEVERAL DAYS LATER, AFTER A DAY with a lot of sailing and not much hiking, I took a walk by myself along the stream we bathed in, heading for the lake from which the stream flowed. Maps and aerial photographs showed that the lake was the most westerly of three that backed into the ice cap, each feeding into the other, each a catchment for water from the melting ice sheet.

The hike took me through small meadows ablaze with white-tufted cotton grass that waved in the breeze like enchanted sentinels. Arctic char from two to three feet long skittered along the bottom of the shallow water, darting from boulder to boulder, trying to hide. Had I been fishing, we would have had a delicious dinner.

The sun filtered through thin clouds, and a light breeze blew. By the time I reached the lake, the day had grown chilly, the lake water choppy. I found a boulder and sat down, jamming gloved hands into jacket pockets, and spent some time in that exquisitely quiet place, watching the lake and fish.

It was hard not to feel overwhelmed by the majestic solitude of my surroundings. Having such moments was a remarkable experience, undisturbed in that profound epitome of untrammeled nature. Life flowed at its own pace, the rock and soil and plants scaffolding a landscape that humans had not shaped. I was the sole audience, a most temporary of visitors, watching the momentary manifestation of flowing processes set in motion from the earliest beginnings of Earth, billions of years ago. What I saw was what that primal force had achieved on its path to future things. Emerging from that sea of possibilities were realizations, ephemeral but concrete, of coincident circumstances, lacking an end point.

For the first time in my life, I felt as though I understood, to the extent I was capable, how utterly incomprehensible that world was for me. Nothing existed separate from any other part of the whole, and the whole was the entirety of the universe, from its very beginning. And there, in the quiet of that Arctic valley, one manifestation of that unity resided.

Time did not exist. The only difference between past and future is the interceding mind, which contemplates and describes and details differences, identifying species, speaking as though they are fixed in time and separate, when, in fact, they are incessantly, furiously changing—temporary, creative, individually unique and yet part of an indivisible whole. Humanity was simply one more experiment conducted by something so immensely incomprehensible that the outcome of the experiment had no importance.

And yet, in that great loneliness, the world was saturated by the beautiful. What surrounded me was stunning in its newness and harmony. Color, texture, form, and pattern flowed from one expression to another without incongruity. There was nothing familiar except the grossest of concepts (rock, water, air, cold); everything challenged comprehension.

Loneliness and cold made it uncomfortable to stay longer. As I stood, I surveyed the scene, trying to capture some pieces of it that I could share with Kai and John, but I realized I did not have the words to convey any of it.

RATHER THAN RETURN TO CAMP along the same track by the stream, I headed cross-country to save time and see new terrain. There was a broad expanse of relatively flat ground that formed an apron around the lake a quarter of a mile wide. It was easy walking and open, one of the few places in that world where I could turn my attention toward something other than where my next footstep should go.

On the way, I came across a field, perhaps two hundred yards long, pimpled by protruding mounds of dirt a few feet across and inches high. They were palsas—small mounds that form when groundwater persistently freezes and expands upward. They are common in permafrost terrains, where pingos (a larger version of the same kind of thing) form. Around the edges were concentrations of boulders pushed up from underground.

I wandered across the tops of the mounds, looking for cracks to see if the underlying ice might be visible, and along the little boulder-strewn valleys at their edges, following a polygonized path through the landscape. Where I walked was like a small maze, a place in which I imagined a mystical tradition of chanting and dancing performed by some unknown spirits, preserved there in that timeless place, patiently waiting for the next generation of believers.

As I walked, something seemed out of place and unusual. Then it struck me: All the boulders were oddly pale in color, with none of the black and speckled patterns or the banding of the gneisses and schists we commonly saw.

The color was due to a lichen of the genus Umbilicaria;

they covered the boulders in a profusion we had not seen anywhere before. Why that should be, I had not a clue. In the tundra surrounding the boulders a profusion of reindeer lichen patterned the ground. The thought then occurred to me that wandering caribou would have had a feast there—it was as though nature had laid out a banquet, allowing an endless indulgence in lichen delicacies. I knew that this would be my chance to find out what, if anything, I had been missing. What did *lichen* taste like?

I carefully removed a small clump of the filigreed, platelike forms from the nearest boulder, cleaned off small grains of sand, and took a bite. The texture was slightly chewy and leathery, but not tough. It was easy to eat. The taste reminded me of a simple white sauce and semolina pasta—nothing extravagant or spicy, just a light, delicate creaminess. There was no great complexity, but a comfortable simplicity that was easy to enjoy. I swallowed the small bite and took another and another, trying to get a better sense of lichen as food.

Suddenly, memories of childhood meals in our small home, tucked next to lemon orchards in Southern California, flooded my mind—the thoughtful flower arrangement that was usually on the table, the patterned tablecloth with faint Early Americana scenes, the milk glass by my right hand, my father to my left, serving us from the casserole in front of him. I stopped chewing and was briefly spellbound in those long-forgotten memories, surprised and disconcerted by faded feelings of childhood comforts. Lichen as time machine.

Could it be that somewhere in my experience of place at that moment, there was a shared element of perception and memory that overlapped with that of the reindeer?

I did not think to try the other lichen. I wish I had, just to get a feel for the world of flavors that encase rocks.

Falcon

I WAS HUDDLED IN THE LEE OF MASSIVE BOULDERS on the summit of a west-running ridge fifteen miles from the edge of the ice. A cold wind blew down from the north, pounding out of the Arctic like an unstoppable train. I was there to collect basic observations that might add some small detail to our emerging story.

The ridge summit along the southern edge of Arfersiorfik Fjord was the highest point of land for miles around. Two yards away, a sharp cliff dropped more than six hundred feet to a massive talus pile at its base. The boulders and rubble that had fallen from the north-facing wall formed a steeply sloping buttress that extended to the edge of the fjord below. To the east and west, the ridge ran for miles, dropping hundreds of feet in elevation from its pinnacle, an undulating bedrock backbone that defined the grain of the land. To the south, for at least sixty miles, unfolded the classic Greenland topography of rolling valleys and ridges, sharp walls and small lakes, carved into a land surface mantled in tundra and littered with boulders, a land skin of wrinkles—furrowed elbows, smile lines, creased brows. A sense of intelligent patience exuded from that skin, conveying the impression of a land that knew much but was content in its silence.

Dark gray clouds hung so low in the sky that I could almost touch their south-rushing underbellies as they pressed a meager layer of rain-swept air against the water-land interface.

To the north, beyond the cliff face, the fjord dominated the scenery, its massive presence defining the place where ocean water and ice melt mingled. I looked down on it and could barely make out the Zodiac in which Kai and John were cruising along the shore, taking measurements—a small dot on a huge, gray, liquid surface, my anchor to humanity. Beyond the water, farther north, the landscape mirrored that to the south.

The ice sheet to the east rested as the inevitable white horizon that dominated the world, a stolid sentry to an ancient land. Even from the height of the ridge summit, I was still thousands of feet lower than the ice sheet's crest. Seven millennia ago, the ice had extended even farther west than where I stood, and everything that I could see was basement to it. Since then, the ice has been in retreat, dropping, as it melted, boulders of all sizes that had been locked in its frozen grip. My shelter from the blasting, cold, wet wind was one of those stones.

Kai and John had dropped me at a point on the shore from which a traverse could be made inland to sample and measure, in order to resolve a detail about whether or not a particular rock unit extended that far west. The answer would let us reconstruct how far one of the faults we were mapping extended. The plan was that I should head due south from the shore, up and over the ridge crest and then down into a major valley beyond. From there, systematically crisscrossing the terrain for about five miles would provide ample opportunity to make my observations. Returning across the ridge and descending to the water's edge, I would rendezvous with them in the late afternoon on the beach of a small cove back up the fjord.

Wandering alone in that infinite, ancient wilderness, planting feet on land that likely had not been touched by human presence, seeing things no other human eye had seen, existing in a world beyond imagination, always discovering something that could not have been anticipated—that, to me, was heaven.

The hike to the ridge crest had been long and arduous. The scramble from the fjord and over the talus pile at the base of the ridge had taken a toll—my shins were bruised and bleeding, my knuckles scraped. The boulders were a field of chaos, some car-sized, some fist-sized, covered in an uneven mat of lichens, mosses, grasses, and flowering plants. That soft, undulating blanket of vegetation, undisturbed for thousands of years, masked thigh-deep holes hidden between the boulders. Firm footing could only be guessed at. If I broke a leg there, it would be late afternoon by the time Kai and John would make it to the designated little bay for our rendezvous and start looking for me. The thought of that long, cold wait encouraged caution. I tried to find small hints as to what might lie below: slight undulations in the dull green surface, the shape and slope of the nearest exposed boulder, the latticework of occasionally exposed caverns—all potential hints where the next best place to set foot might be. Even so, all that attention remained little more than guesswork. Inevitably, a few bounding leaps from boulder to boulder would be followed by a sudden crash into a covered hole, a struggle to climb out, moments spent rubbing bruised and scraped shins, a few deep breaths taken, and then the need to push on. There wasn't time to do anything else.

The feel of the mosses, though, was unforgettable. At first, I wore gloves, so I missed the sensation. But about halfway up the talus slope, when I had fallen in yet another thigh-deep hole, I decided to rest for a minute to catch my breath. At exactly eye height, a rock two feet across and directly in front of me beckoned. Moss draped over it like a shroud and flowed onto the surface of the surrounding boulder field. Below that stone was a small cavern exposing the underside of the vegetated boulder. The combination of the black-and-white lithic form, the velvety green texture of the moss, and the coolness of the air inspired me to pull off a glove and brush my hand over it. The feel was stunningly luxurious, as though the world's finest velvet, a foot thick and plush, had been arranged over the boulders in unabashed, delicate extravagance. Climbing out of the hole and walking on, but with gloves off, it was hard not to feel guilty trampling on something so beautifully offered.

THE TALUS MET THE ROCK WALL at an elevation of about nine hundred feet. A bare face rose out of the talus jumble and stepped up to the crest of the ridge in steep, short pitches. That part of the climb was relatively easy and got me to the top quickly.

By the time I attained the ridge crest, I assumed it was around noon. I quickly had lunch —canned sardines, cheese and stale rye bread, raisins and chocolate, and some water. The boulders stood imposingly, littering the ice-polished rock. My nose ran and my eyes watered in the blasting, freezing wind. Stones had to be placed on anything I took out of the backpack so it wouldn't blow across the ridgetop and out into the valleys.

Once lunch was finished and the backpack reorganized, I walked over toward the cliff edge. I wanted to stand in the raging wind, gaze out at the endless view, and feel the wildness of it—pure in its coldness, present as an absence of everything. I held out my arms to let wind pound every part of me. But the cold was overwhelming. I dropped my arms, stuffed gloved hands into anorak pockets, and just looked out on the vastness.

For a few moments, nothing disturbed the absolute, stolid permanence of the land. Despite the roaring wind, what could be seen was a passive world, rock-hard, still, and unmoving. Then, just at the edge of vision, off in the direction of the white ice sheet, a small, dark, incongruous dot moved. I slightly turned my head to see if it was real.

I had a moment's difficulty fixing on it, but it quickly

resolved into an almost invisible black speck moving just above the ridge crest, riding the wild updraft of the wind streaming off the rock wall. The speck was moving very fast and rising toward me. Before I had much of a chance to think, it was at my elevation and closing in like a rocket. Its trajectory would take it within a few feet of my head.

In a flash, I realized it was a small peregrine falcon, wings held tight against its body, taut, attentive, barely more than a feathered projectile—aerodynamic perfection riding the invisible streamlines washing over the ridge. Its wings barely moved, adjusting ever so slightly to keep it a few feet from the cliff edge with every change in the speed of the wind.

When collision seemed inevitable, I took a step back to get out of its way. Suddenly, time shifted, as it sometimes does when the unexpected shocks us. Every movement and motion, every thought and sense distilled into crystalline clarity. Seconds and fractions of seconds lengthened. What I saw was remarkably sharp.

The bird flared its wings, reared its head, its dark eyes opened wide. From less than thirty feet, it glared at me, seemingly suspended and utterly motionless in midair.

Then, with subtle grace, it tucked its wings back against its strong, elegant form, slightly changed course, and sped away on the wind. Twice, as it flew toward its unknown destination, it turned its head and looked back over its shoulder, as though trying to assure itself that what it had thought was there on that ridge crest was, in fact, real. The sound of its feathers against the wind was a muffled, swooshing hiss.

It is impossible to know what that bird experienced during our brief confrontation. It is likely that its flight through the blasting updraft had been one of concentrated attention to wind speeds, gauging distances between itself and the rock face, flying toward some far destination, the scattered boulders along the ridge crest simply impressions of landscape and shadow flashing by, until one of the boulders moved. High on that ridge crest, a human being was not expected.

To meet that animal at such close range, unintentionally, innocently, was something impossible to imagine in any other setting. Thrilled excitement and shock pounded through me as I realized I had just experienced the purest statement I would ever know of what it means to be wild.

WHAT WE EXPERIENCE MUST BE SEEN as an altered reality, a tinted fragment. Everything new, whether a physical place or a cognitive construct—a landscape, birdsong, a blanket of mosses—becomes associated with names and emotional impressions based on what we remember. Through that process, it becomes its own memory, against which we compare the next experience. The implication is clear: The richer the past that is contained in memory, the stronger the congruity with the moment will be, and the better we will know what the world is.

Is everything relative, then? Are all the experiences I reflect on and wonder about a simplified collage of what I have seen and felt? If so, what I can imagine is limited because of the boundaries of my past. Every new experience that fits no previous memory is a gift that enriches the vault of color, sounds, and smells, the wealth of emotions and depth of meanings I have available to me. What is new embellishes all future experiences.

Wilderness, through the fact of its existence, is new.

IMPRESSIONS II

In the face of a rational, scientific approach to the land, which is more widely sanctioned, esoteric insights and speculations are frequently overshadowed, and what is lost is profound. The land is like poetry: it is inexplicably coherent, it is transcendent in its meaning, and it has the power to elevate a consideration of human life.

—Barry Lopez

We are the result of water insinuating itself into the latticework of crystalline forms, of its persuasive discourse with the elements that reside there to run to the sea. Water encourages unities and pairings; it facilitates the necessity of elements to become molecules, and molecules to form the most complex construct the moment might allow. But water, too, is the catalyst for decay and dissolution. Water decomposes rock just as surely as it encourages reconstruction.

It is that process of relentless reconstruction that made us. We live in an illusion that is a consequence of our trial-and-error biology. Our reality, consequently, is an impoverished truth. In pristine wildness, one has a chance to experience small epiphanies that expose one's preconceptions and misunderstandings.

CONSOLIDATION

When we try to pick out anything by itself,
we find it hitched to everything else
in the universe.

—John Muir

The Sun Wall

To the south of camp and just a bit east stood a majestic rock wall. It rose out of the water, a massive buttress around which the fjord jogged southeast for several miles before continuing its eastward trend to the inland ice cap. It stood nearly a thousand feet above sea level, and dominated the world where we camped.

In summer, the sun makes a lazy circuit in the sky, neither setting in the north, even at midnight, nor rising more than forty degrees above the southern horizon when it reaches its apogee at noon. Low sun angles make for striking shadows, and with the sun orbiting the entire sky, the face and form of things never stay the same.

My tent opened to the west, giving me a view down many miles of open fjord water. But the rock wall was to my left and behind me when I would emerge in the morning from my tent. I always turned toward that massif to get a sense of what the day's weather might be. It was, of course, no way to gauge how the day would evolve—high Arctic weather is notoriously mercurial, but even so, somehow seeing that bulwark in the morning light gave a perspective on the day that would sit with me until we got back to camp in the evening.

On a clear day, the morning sun would be low in the northeast, hanging just above the white ice, a few hours into its slow ascent in the sky. The rock wall, as a consequence, was backlit and resting in shadow, dark, flat and mainly featureless. Blue sky would blaze behind it, with even bluer water stretching toward me.

By the time high noon came, details boldly stood out in the oblique light. Chimneys, chutes, ledges, and overhangs, all cast in shadows of varying depth, became prominent, adding texture to the surface that was missing in the morning. As the afternoon progressed to evening, the positions of shadows shifted, and their scale and extent evolved. Color emerged in the rock face indicating that plants lived there, tenaciously clinging with pervasive roots to cracks and seams. Tundra-filled valleys that lapped onto the flanks of the ridge took on hues of rust and sand mixed with the greens and grays of the leafy foliage.

It was hard not to think of that entire scene as a canvas upon which the sun painted, ceaselessly retouching every inch.

THE SUN DID NOT ALWAYS SHINE. One morning, when we had planned a long trip down the fjord, we came out and found a dense set of broken clouds hanging over us. A chill wind blew hard, and the waters were choppy. We revised our plans accordingly and traveled up a small inlet near camp and worked that geology in detail. It was a segment of the northern edge of the shear zone we had not seen.

In a tiny bay off the minor fjord, Kai noticed an unusual sequence of off-white and deep green bands running in an exposure just above the high-tide line. John carefully nudged the Zodiac up the shallow reach to the water's edge and grounded it. After securing a line to a large boulder, we walked the shore over to the outcrop Kai had pointed to. Surprisingly, we found thin layers of marble, sillimanite schists and rocks rich with carbonate and silicate minerals, all the likely signature of shallow-water sediments deposited along quiet shores where microscopic, single-celled life had thrived in warm oceans. Had we been there at the time tides were washing those shores billions of years earlier, we probably would have been swimming in the shimmering, crystal-clear waters of a small bay.

Now, recrystallized as a consequence of being deeply buried and cooked at hundreds of degrees in the Earth's interior, the limestone had become marble, and the muds and sands had been transformed to green gneisses and schists. How deeply buried they had been, we could not tell, but the minerals we could identify could not have formed had the rocks not been buried at least ten miles down. We stood on more evidence of an ocean, a consistent element that would be expected in an area where a suture might be.

About midday, the skies cleared and the wind died down. We relaxed somewhat in the warmer air, eventually heading back to camp in the late afternoon, pleased with what had been accomplished.

When we got there, we anchored the inflatable, unloaded

samples and gear, and headed up to the cooking tent, where we spent what was left of the afternoon compiling notes. John sat on one side of the tent, reading through some of the papers he had brought with him, making notes in their margins. I was across from him, rewriting the scribbles in my field book so that they would be decipherable— my handwriting has never been good. Kai was near the entrance to the tent, preparing dinner. Onions and butter sizzled in a pan on the Primus.

The data we had collected were increasingly supporting the notion that the region preserved a record of intense deformation, as John and Kai and others had originally argued. The pencil gneisses John had first found in that one outcrop near camp and which were irrefutable evidence of extraordinary shearing at high temperatures, turned out to be a common feature for miles along the shear zone. The lenses of pillow basalts and ultramafics, too, were likely proof that hundreds or thousands of miles of ancient ocean floor had been dismembered and sliced into thin bodies, a process requiring staggering amounts of displacement and deformation. And all this was localized within the shear zone.

But emerging, as well, was much more complexity than had been expected. Although certainly a zone of deformation, the remnant slices of seafloor required processes that could consume entire ocean basins, leaving little more than the slivers we found. The presence, too, of sediments like the ones we had seen in that small bay a few hours earlier suggested the edge of a continent.

Not more than a mile from where we sat were portions of the magmatic remnants of the Andes-like volcanoes that had implied the existence of a zone in which ocean crust was consumed. Taken together, the simplest concept that could explain all these observations was that our camp, serendipitously, sat in the collision zone Kalsbeek and his coworkers had postulated. If so, the shear zone John and Kai had worked on was a much more profound tectonic feature than anyone had imagined; it was the actual suture that locked together continents that had collided 1,800 million years ago. None of that was part of what they had discussed in their earlier work.

I BECAME A GEOLOGIST BY ACCIDENT. When I was growing up on the Southern California coast, surfing had consumed my life. In high school, I missed classes to surf, nearly failing many of them. I sat in detention rooms and was expelled several times, but the call of the waves always lured me away from class. I could not resist the temptation to immerse myself in the thoughtless uncertainty each wave offered. Sitting on the board, anticipating the next swell—the opportunities for success or failure, the anxious adventure of not knowing the outcome of a self-imposed dare—there was nothing better.

When the time came to pursue education beyond high school, I chose a college that was farther south along the coast and that offered courses in oceanography. I was convinced that I could dabble in that science as a career while committing most of my time to bottom turns, hanging ten, and getting locked in.

But at that university, oceanography was a graduate-level pursuit; an undergraduate interested in oceanography had to major in biology, chemistry, geology, or physics before applying the principles of the chosen discipline to study of the ocean. I reluctantly chose geology.

I suffered through one course and took another, only slightly interested in the subject. Then one day on a required field trip, the professor teaching the course pulled the van over to a small outcrop at an unplanned stop. I suspect he had sensed boredom in the air. He got us out of the van and gathered the group around him.

"I want to show you what we are training you to do," he said, and then pointed at a black mineral in the crystalline face of the road cut. For some minutes he described the significance of the mineral, named it, and explained its chemical composition. He pointed to another and did the same thing. After five minerals, he wove a tale that surprised us all. Where we were standing had been the middle of a chamber of molten rock 65 million years ago, ten miles below the surface. He went on to tell how it had formed, what volcanoes it had fed, what its history had been once it cooled to a frozen state. I was mesmerized. Suddenly, I understood Earth to be a manuscript, written in an extraordinary calligraphy, embellished with an artistry I could barely discern. Mysteries of immense proportion, histories of our origins, and the collection of accidents that had made us what we were resided in stone everywhere. In an instant, the world had become a new place for me.

A WARM FOEHN WIND GENTLY BLEW from the east, descending the thousands of feet from the inland ice "summit," lightly ruffling the canvas of the tent as we talked. The low sunlight from the west lit its orange fabric, infusing our little room with a warm glow.

Then, without a hint that change was coming, the light suddenly dimmed and the wind stopped. We joked with one another as the tent cooled. The wind began to blow from the west, gently at first, sending the tent canvas into small but insistent convulsions. Then, within barely three minutes, the wind picked up and we were buffeted by pounding gusts. The tent flaps snapped and the walls bowed in, pushing down on our heads. Kai turned down the Primus; we stopped our talking, dropped our notebooks and pens, and ran outside to see what was happening.

The fjord had turned into a dark gray maelstrom of blown-out whitecaps and east-running waves. Long white streaks of foam formed perfectly straight lines on the choppy surface. Wind tore at everything, howling with gale force; we had to lean into it to stand up.

I shifted my eyes from the water to the rock wall that I scanned every morning. There, an epic battle was taking place, the like of which I had never seen.

The gale was screaming down the fjord from the

west, directly into the massive stone buttress. The wind slammed head-on into the wall and had nowhere to go but straight up. As it did, streamers of clouds condensed out of thin air, forming vertical white stripes hundreds of feet long, rushing up and over the rock face, decorating it in speeding, billowing ribbons. When the wind and clouds reached the summit, they ran off to the east. Cloud fingers miles long angled upward from the crest of the wall, racing at incredible speed toward the inland ice.

Suddenly, I heard John yell in a frantic voice, "The boat!"

I turned toward the little cove that we had anchored in and saw a disaster unfolding.

John had devised an ingenious anchoring system for dealing with the huge tidal range there. Normally, where the tides were small, we would have simply dragged the boat up the beach to a spot above the high-tide line and tied it off without fear of losing it. But there, that was not possible. The twelve-foot tidal range overwhelmed the span of beach. So John had dropped an anchor about a hundred feet offshore, and tied a buoy to it. He had also tied a pulley to the buoy as well as one to a rock onshore. A rope looped through both pulleys allowed us to secure the stem and stern of the boat and then haul on the ropes until the boat was some distance offshore; in the morning, we would simply pull the boat back in. That way, regardless of the tidal stage, the boat could ride out the ebb and flow, safe from the rocks onshore.

From our camp, though, we could see the boat caught

by the gale wind, dragging the anchor in a large arc toward shore. It was headed toward a promontory of jagged rocks. We had patching supplies, but not enough to repair the entire boat if it were torn up, and we had no backup. We desperately needed the boat to finish our work; without it, we would be helpless, the summer would be lost, and we wouldn't have another chance to return for more than a year. Our only chance was to get the boat ashore, and there was precious little time for that.

John was already in a flat-out run for the cobbled beach. Kai and I took off after him. We reached the small cliff above the beach at the same time and took turns scrambling down it. John sprinted across the cobbles and grabbed the line to the boat. The three of us began hauling on the rope. For some obscure reason, locked in the physical geometry of our situation, every time we pulled on the rope, the boat accelerated in its race toward the rocks. We stopped for a moment, trying to figure out how to solve the problem as the boat moved toward its destruction. We only had seconds, but there appeared to be no answer other than to pull. It seemed hopeless, since we could clearly see we did not have time to haul in enough rope to keep the boat off the rocks.

"There's no choice. We have to pull!" Kai shouted.

Without hope, we grabbed the rope again and pulled.

We struggled for some seconds, pulling as hard and fast as possible, all the time watching the seemingly inevitable disaster unfold. Then, when the boat was within just a few feet of the first jagged rock, the wind suddenly dropped to almost nothing. The boat stopped moving and began lazily drifting with the tide back toward the buoy. Within minutes, the gale was over, the foehn wind began again, and the sun came back out.

Relieved, John reset the anchor and Kai and I headed back to camp. As we walked, I turned to look at the stone buttress. It was bathed in shafts of cloud-broken sunlight. Although shadows came and went, in that late-afternoon light the cliff face glowed.

Bird Cries and Myths

WE WERE WORKING THE SOUTH SHORE of Arfersiorfik Fjord, far to the west of the pillow basalts and the rock that smelled like singed hair, searching for more evidence of that old ocean floor. The location was about as far west as we could sail and still have time and fuel to make it back to camp.

The day was clear, the wind light and out of the north, the temperature warmer than normal. It had been a long but productive morning—a few good samples collected, some measurements of the rock fabric noted in our field books, even a few hints that there might be the kind of metamorphic history we were looking for. We decided to take a break and stop for a quick lunch in a small rock-protected notch along the shore. John turned the Zodiac toward the beach and gunned the engine briefly before shutting it down and pulling up the prop. Kai and I jumped out when the boat lurched onto the sandy gravel, then dragged it up above the reach of the rising tide and tied it fast.

We found a small pocket of sun-warmed rock, threw down our backpacks and settled in for lunch. Eating our kippers and rye bread, and drinking coffee from the thermos, we talked about what we had seen and what to visit farther west along the coast. As the debate evolved, an afternoon wind began to blow, coming strongly out of the northeast. Chop formed on the water, blowing against the tack we would have to sail to get back to camp. Beating against it would make for a hard trip, so we decided to cross quickly to the northern shore, where we could sail in the lee of the hills and bluffs. Since there was a lot of geology along that coast that had not yet been mapped, that plan had the benefit of letting us fill in a few more data points on our sparsely populated field map. We cut our lunch short and repacked our backpacks, quickly scrambled down the scree to the graveled beach, and pushed off.

The first site of interest on the north shore was an enigmatic white blotch on the aerial photographs we used to plan our work. In the old print, taken from twenty thousand feet decades earlier, a featureless white area about half a mile wide, just inland from the north shore and separated from the water by a narrow peninsula, sat like a blemish on the image. It appeared to have a small inlet from the fjord, and sharp cliffs surrounding it, but otherwise, there was nothing to suggest what it was. Its whiteness stood out in sharp contrast to the muted grays and blacks of the tundra, water, and gneiss that surrounded it.

Because there was a flooding tide, John set us on a heading that would intersect the north shore about half a mile west of the inlet, which would allow us to cruise slowly back with the incoming tidal flow.

The trip across the two miles of open water, sailing

against a five-knot current, took about twenty minutes. We kept looking for the whiteness we knew had to be on that far shore, but a small bluff hid it from view. Once we reached the other side, John swung the Zodiac around and we slowly rode the tide east along the water's edge, our anticipation growing with each minute, as we wondered what that white land would turn out to be.

Within a few hundred feet of the inlet, the bluff hiding the white terrain dropped away and we caught our first glimpse. Right at the water's edge, a flat shelf of bedrock gneiss tens of yards wide and perhaps fifty yards long made a little shoal just a foot above the slowly rising fjord waters. The gneiss was washed clean by the endless rising and falling of the salt water over it. The aerial photo of this small peninsula must have been taken at a much lower tidal stage. John ran the Zodiac right onto it.

The featureless white area turned out to be a vast tidal flat of very fine mud, surrounded by a broad white beach that ended in sharp cliffs of pure white powdery sand and silt. The cliffs were cut into sediments deposited by streams gushing from the base of the ice sheet thousands of years ago. From the look of the sediments—the fore-set beds capped by a few meters of flat white silt—those ancient flowing waters must have formed broad deltas as they flooded into the frigid fjord. Although the present ice front was nearly forty miles east from where we were, the edge of the ice must have been less than a mile away when the white sediments were first deposited as deltas. The tidal flat was made of those same sediments, reworked and redeposited by ebbing and flooding tides and seasonal rains during the thousands of years since the ice had retreated.

The white surface of the tidal flat was completely barren of plants—a rare Arctic desert of fine mud existing as a sterile membrane protected by the bedrock rim at the edge of the fjord. When it was exposed to the air at low tide, the mud dried slightly, becoming a featureless off-white presence. It quickly became apparent why no plants grew here—tidal cycles kept the mud brackish, and the glacial outwash sediments were devoid of nutrients.

I walked across the small peninsula to the edge of the sediments and knelt down. The surface was featureless and almost perfectly horizontal, the smoothest and flattest natural expanse of land one could imagine. Fjord waters a fraction of an inch deep flowed over it, slowly making an ingress as the tide rose. The afternoon light shimmered on the mirrored film of water, reflecting a pale sky and white cliffs.

With nothing to hunt or gather there, the likelihood was small that anyone had ever disturbed that place. The aura it exuded was of a barrenness that seemed to mirror much older times, billions of years ago, when there were no land plants, when the hills and valleys and rolling plains of the earliest Earth were rock and blowing sand. It would have been a place in which life was able to survive only if saturated and immersed, covered by liquid, impregnating mud.

I walked along the bedrock shore and around the

muds that were drying in the sun. Glistening, succulent, resting in a color just this side of white, the mud was irresistible. Kneeling down and slowly pushing my fingers into it, I wondered how deep it might be.

Weirdly, I could see my fingers penetrate the top fraction of an inch, but the muck was so fine and so water-saturated, so perfectly balanced with the air temperature that there was no resistance or sensation. As I plunged my arm deeper, it was like pushing through a magical wall into a different realm, a place where everything was alien and imaginary.

Half an inch below the gray-white surface of clays, a fluid, organic black ooze glistened on my fingers. With the protective membrane of clay broken, the underlying, thriving biology flooded the air with the sulfurous aroma of its complex, primitive world.

Three billion years ago, communities of single-celled life colonized the tidal pools and flats of early Earth. Living things ran riot, unencumbered by anything except the limits of nutrients and the boundaries of water. Surface clays in tidal flats protected fragile organic molecules from ionizing ultraviolet radiation, and held in the moisture necessary for life's chemistry. Tidal cycles replenished what was diminished; sunlight kept it warm. Resting there in that silent mud was what we came from, our vestigial home.

John and Kai were a quarter of a mile away, measuring the strike and dip of the colored layers that composed the bedrock gneiss, establishing the fabric that long postdated that birthing process. Eventually, I walked over to them, hands dripping with rapidly drying mud.

Just then, Kai turned toward the Zodiac and said, "Gentlemen, we need to hurry. The tide is lifting the boat." I grabbed my hammer, pounded off one more sample, which I quickly labeled and bagged, and then ran to the boat.

As we powered out into the fjord, we suddenly heard a strange sonorous wail over the whine of the engine. Unable to make sense of it, we ignored it at first. But the strangeness grew and persisted, eventually causing John to power down the outboard so we could listen.

The sound came from across the fjord, more than two miles away, mournful, wrenching, and melodic. As we listened, it slowly morphed into a feminine symphonic chorus.

We decided it would be irresponsible not to investigate—perhaps a small fishing boat had sunk and people were stranded, or some other tragedy had happened. John swung the bow around and we started back across the fjord.

Within a very short distance, the cries changed. At first, the wailing became fragmented and less sonorous; then, we heard bursts of sound and staccato screeches. John stopped the boat and we listened again.

The southern shore of the fjord was a massive rock wall, rising hundreds of feet straight out of the water. It was barely in shadow and its face was patterned grays. At first, that was all we saw. Then, after straining our eyes, we spotted hundreds of circling gulls wheeling on the updrafts coming off the cliff. The rock face was, in fact, a rookery. From their numbers and crying, it seemed something had startled the birds, perhaps an Arctic fox, perhaps the noise of our outboard—it was impossible to know.

Amused at being duped, we headed back the way we had come. As we made it to our original position and course, the wailing began again, the bird cries morphing back into rending cries of distress.

WHAT WE HAD EXPERIENCED could be easily explained. Over the cold fjord waters, chilled air accumulates, forming a dense layer perhaps a few feet thick. The air at higher elevations is warmer and less dense. Since the speed of sound varies with air temperature and density, sound waves become distorted and pitch changes when the waves are refracted through the stratified fjord atmosphere. In most places, the effects are minimal and not even noticeable—words spoken in one place will be heard as intended. But if the conditions are right, the refraction can be dramatic, the sound distorted. Sitting in our small boat, our ears immersed in the chilled, dense air, more than a mile from where the birds uttered their cries, the propagating sounds stretched into an aural mirage.

Such an explanation, though, trivializes that experience. Cruising along the edge of the fjord, heading back toward camp, it came to mind that what we had heard was almost certainly the sound of the Sirens, the mythical creatures Odysseus heard more than 3,200 years ago, tied to the mast of his ship as his men toiled to keep it on course, wax in their ears so they would not be lured to their destruction by the Sirens' song.

We had entered that place below the surface of things where nature fosters the birth of myth. Our little diversion into the fjord was an excursion across a permeable membrane.

Ptarmigan

INEVITABLY, TO COEXIST AMICABLY with those one came with into the wild, one must bathe. While certainly bracing, bathing in the field in the Arctic is a duty, not a pleasure. There are two reasons why this is so. One is that most streams and lakes are ice-fed, making the water very, very cold. The other is that on a clear, sunny day when there is no breeze and temperatures are appealing enough to make bathing attractive, clouds of mosquitoes descend by the hundreds, if not the thousands, to gorge on naked flesh. The only solution is to bathe when there is a breeze strong enough to keep the mosquitoes downwind, which makes immersion in the water unimaginably painful.

On a particular day in July, when the sky was gray and a slight breeze blew, it was time. It had been days since I last bathed, and I was ripe. After a few hours spent steeling myself that morning, waiting for the temperature to creep up just a degree or two more, I grabbed soap and a towel and headed off.

The stream where the Arctic char swam was to the east of us, a little more than a quarter of a mile away. It tumbled in a torrent through a small boulder-choked gully just before entering the fjord. The rush of water was fed from a chain of three lakes, the most easterly of which sat at the immediate edge of the ice sheet. The walk to the stream was an easy stroll, done in a few dread-filled minutes.

Arriving at the stream, I walked along, looking for a small sheltered pool. Rather sooner than hoped, a perfect spot emerged from around a small bend. Water fell into a small catchment deep enough to submerge myself in, providing just enough space to duck under the frigid cascade.

Taking a deep breath, I quickly undressed and plunged in. To say that it took my breath away is an understatement—the gasp that escaped my lips was probably heard back at camp. A sharp, stinging wave of intensely burning cold exploded from every inch of skin as I shuddered and writhed. As quickly as possible, I soaked myself, stood in the wind and lathered up, then dived again under the waterfall to rinse off. The total amount of time spent in the water was probably less than three minutes, but it felt like hours.

Scrambling out of the water and standing precariously on wobbling boulders, I dried as fast as possible in the biting breeze. My skin was red and burned from the cold, and the scratchy towel seemed to do little more than smear the water over goose-pimpled flesh. I stumbled over boulders, stubbing my toes, to where my clean clothes lay in some bushes and then put them on as an aching numbness began to affect my feet and hands. Once my clothes were on, the relief from the chilling wind was exquisite.

The walk back to camp first led along the pebble beach at the mouth of the stream, then up a small bluff to the tundra bench. As I ambled along, the feel of clean skin under layers of insulating clothes was invigorating. The harsh prickling from the cold subsided and a fresh sense of sharp sensitivity to light, air, and smell gave an impression that the world was somehow newly refreshed. Everything seemed vibrant and intensely real.

Lost in thought, walking through the grasses and short-stemmed flowers of the tundra carpet, I experienced a feeling of belonging, of being in a welcoming spaciousness, a sensation that replaced the dread that had accompanied the bath. I relaxed and could feel my muscles loosen.

Then, off to my left something flickered at the edge of sight. I ignored it for a few steps, not wanting to interrupt the simple pleasure of walking through that quiet place. But fearing that I might be missing something, I stopped, turned around, and took a few steps back the way I had come. Suddenly, as if materializing out of nowhere, a female ptarmigan about the size of a large chicken quickly scuttled away, just five feet from me. She didn't move far, perhaps a foot or two, before she settled back into the tundra and puffed up her feathers. Despite the short distance between us, finding her in the tundra took intense concentration. The patterns of brown, tan, and black markings on her matched precisely the pattern of color and texture of the plants she was nestled in. I stood mesmerized by the visual magic she performed, leaning my head to one side and then the other, trying to find some position that would allow me to see her, but she insistently melted into the scenery.

To gain a different perspective, I took one step to my left, and then something else moved. Three feet behind her, a tiny hatchling darted away, only to stop among the foliage and vanish. And then, almost next to that first hatchling, another baby ptarmigan briefly materialized, huddled down in the vegetation, it, too, nearly invisible. Not wanting to scare them any more, I backed off a step, but that only startled the mother again. She ran toward the two hatchlings, and together, the three of them froze. Astonishingly, another tiny bird appeared, exactly where the mother had been standing; she had been protecting it under her wings, until she could not take the tension any longer. I backed away a few more feet and tried to see if there were any more.

I knelt down on hands and knees and then lay on my belly to see if the form of the little birds could be seen against the sky, hoping I might discover even more of them. As my face came within inches of the ground, I was suddenly awash in layers of sweet flower scents. As I rested lightly on the surface, the smell of dozens of blossoms I hadn't noticed engulfed me. Arctic poppy and white Arctic bell-heather were interspersed among mountain sorrel, hairy lousewort, purple saxifrage, and mountain avens. I was awash in a botanical sea, carried into an unexpected world.

For a moment, the birds were forgotten and my attention focused on identifying individual scents from different flower species, but the complex mix of fragrances eluded me. The smells came and went, as though wafting over the ground in waves and streams, riding the will of the gentle, wandering breeze. No wonder the bumblebees almost always zoomed along at ground level, buzzing their way among the flowers there. Scent was their map, and that map rested just above the vegetative surface. Olfactory pleasures flowed there, each component of the currents marking the presence of a sought-after flower. The organic signatures, sensed by us as odors, must have been more than that for the bees, but what?

Happily covered by the smells of the flowers, I looked again for the birds and spotted one more a little farther off, behind the first two that had made themselves known. The mother loomed over her brood, doing all she could to protect them. Finally, probably as a last, desperate move, she limped off, faking a broken wing and injuries, trying to lead away the hulking human intruder.

Riddled by guilt for having stumbled into her life as a perceived threat, I started to leave, realizing those small birds knew the world as I never would. Within a few inches of the surface, the breezes we are familiar with become tempered by stones and boulders and tundra hummocks. In such stillness, scents can accumulate and mingle. A world of perfumes would cloak the hatchlings and saturate their feathers, becoming a sensory background to the birds' accumulating experience of living, the only reality they would know.

As I stood up, the scents vanished. I breathed deeply, searching for a hint of them, but there was nothing more to smell in the air I walked through.

THE OBVIOUS LESSON WAS that scale matters. This world is not designed for us; we populate and experience a very small part of it. We evolved to fit optimally, more or less, within a certain volume of space less than eight feet high and a few feet wide. We do that well. But we are not ordinarily privy to the world that exists within the tangle of tundra plants and saturated soils, nor the complex of forms below the tidal range, nor the chaos of currents the falcon flows through. Not paying attention to these things leaves us impoverished and ignorant.

To some extent, science provides access. It attempts to delve below the surface experience and provide descriptions of what exists there. Regardless of the scale examined, the pursuits of science have shown that there is much more within each realm than imagination could ever, ever produce.

What science cannot provide or explain, however, are the human experiences those spaces inspire, nor why we pursue understanding them in the first place. Knowing the mathematical and objective description of place only feeds the hunger to understand, while that hunger remains one of the greatest of all mysteries.

Clear Water

THE BEDROCK, BACKBONE TO THE LANDSCAPE, shapes impressions and guides the winds. Tidal flow is constrained by it; the ice rests on it. It is impenetrable. We break off samples with ringing hammers and nothing flows out, but in that crystalline scaffolding, water resides. The water is inherited from the time the rocks were little more than mud sludge on the ocean floor. Slowly buried and recrystallized, the atomic lattices of evolving new minerals capture the water molecules in systematic arrangements, preserving them for future considerations.

GREENLAND IS INCISED BY THOUSANDS OF FJORDS and fringed by countless islands and skerries, giving it a length of coastline as great as the circumference of Earth. The ice sheet that dominates it contains more than 600,000 cubic miles of frozen water. As a consequence, it is a place defined by water. As one becomes sensitive to that reality, unexpected perspectives present themselves. The extent to which water and rock are consanguineous must be addressed.

FAR WEST OF THE ICE CAP, THE FJORD WATERS are free
of silt and mud and are crystal clear. Many years before,
when I left for my first summer expedition in Greenland,
I already knew it would be a place dominated by the sea.
Knowing that intellectually and experiencing it in reality,
however, are two utterly different things.

One early afternoon on an unusually warm day on
that first expedition, walking along a low ridge that was
the heart of a small peninsula, I glanced down to my left
toward a little bay of crystal clear water that ended at a
stony beach about a quarter of a mile away. The nearly
vertical rock walls that bounded the north and south
sides of the bay plunged directly to its floor. The water
was about fifteen feet deep. Sunlight lit the seafloor so
brilliantly that the colors of the underwater world, which
are usually muted and dim, pierced the air in a vibrant
shimmer. Every imaginable shade of green and purple
and gray, highlighted by splotches of yellows and blues,
glowed there.

Incongruously, a nearly black gash asserted itself, riding in the water right next to the shore below me. Three
feet long, linear, and near the surface, it stood out in
sharp contrast to the asymmetry of the free-form backdrop of dancing colors on the seafloor. Slowly, it moved
inland toward the beach, seeming to drift with the tide.
At first, I thought it was a piece of driftwood, riding the
currents from some faraway place where wood grew
and logs existed. Eventually, through exquisitely subtle,

swaying undulations, it revealed itself to be a fish, slowly swimming in that crystalline liquid space. It did not seem hungry, or on the prowl, more as though it were relaxing in the high noon light, casually taking in the serene tranquility of its world.

Later that day, walking back to camp, the wonder of that liquid space haunted me, and I decided I needed to see more of it. We had a small skiff for short excursions to sample and explore the cliff faces bounding the bays near camp.

With fifty yards of monofilament line, a single hook, and a two-ounce lead sinker, I rowed across the inlet where our camp sat, making for a cliff face across from us that indicated it might be a good fishing spot. It was late afternoon; sunlight slanted at a low angle, illuminating the rock wall with a water-filtered elegance. I stopped rowing and looked straight down in the crystal clear water, curious at what could be seen of the bottom. But everything there—the seaweed-encrusted stones, fish, shellfish, and cobbled seafloor—shimmered and flowed, causing a feeling of vertigo. Something seemed to manipulate the light, unnatural and unexpected.

The bay was the outlet of a small stream that ran behind our camp. Water babbled over stones and wended through grassy stretches, soaking up what warmth it could from the sun and land surface. The water of the bay was icy cold. When the stream entered the sea, it floated as a freshwater tongue on the cool density of salt water. The result was a layer of fresh water several inches deep flowing across the bay on the back of the sea. The interface between the fresh water and the salt water was a boundary of contrasting densities, mixing in small gyres and tiny internal waves. The difference in temperatures and compositions of the fluid masses bent the light reflected from the bottom, distorting the patterns, twisting the colors.

I reached over the side and put my fingers into the freshwater. As they moved down a few inches, my fingers penetrated the slithering boundary layer. Painlessly, I watched as flesh disassembled into a dance of swirling abstractions, my fingers becoming nothing I knew.

As I pulled my fingers out of the water and continued rowing toward the cliff face, an aura of enchantment settled over the bay, as though I was entering a world that had been waiting to be seen. Above the waterline was a series of bands of rusty browns and white, typical of the sulfide-rich gneisses and schists in the area. Approaching the cliff face, though, it became evident that the colors below the waterline bore no resemblance at all to that pattern. The waterline became a discontinuity, severing the submerged world from the land surface with impressive precision. Underwater, no hint existed of the obvious banding on land. Instead, a rich deep purple covered the wall. The water was at least thirty feet deep and, although the clearly visible bottom was a random mix of light-colored boulders, sand, and gravel, the entire wall was a simple dramatic purple from waterline to bottom.

Only when I was a few feet from the underwater cliff face did the purple resolve into thousands of sea urchins, so densely crowded that their spines tangled together in an organic, spiked weaving. Barely an inch of space existed between any of them for hundreds of feet. Looking closer, it became clear that what seemed a static purple surface writhed in subtle motion, each urchin slowly making its way through that forest of individuals, spines lazily waving in the current, grazing on whatever algal remnant had been missed by its neighbors. For some minutes, I drifted along the water's edge, marveling at the nature of urchin existence, a biological complexity lacking mind but driven by the urgency to eat.

Eventually, I pushed away from the wall to examine more of that submarine tableau. My eyes, though focused on the seafloor thirty feet below, were caught by something out of focus and just below the surface. At first, what seemed to be floating in the water was an iridescent wire slightly and repeatedly rippling in invisible waves. Then, as though a veil had suddenly been lifted, that one wire resolved into a collection of hundreds, moving in a slow ballet with the gentle current. I pulled in the oars and leaned over the gunwale, trying to make out what this was. "It" turned out to be hundreds of small comb jellies—marine invertebrates that look like jellyfish but belong to the phylum Ctenophora (jellyfish belong to the phylum Cnidaria). Each was shaped like a lantern three to four inches long and two inches across. Along the length of each body ran eight thin threads of cilia that glowed with iridescent colors as the cilia propelled the slowly turning lanterns in the sea. The cilia beat in rhythmic waves that flowed along the nearly transparent bodies, giving the impression of thin threads of rainbow colors tumbling in the clear water. They surrounded the boat as far as I could see, immersing me in a world of shimmering kinematic magic.

There was nothing to do but relinquish intent and float with the jellies. I lay down in the boat, head braced on the stern board, and gazed at the silent spectacle of light and colors, mesmerized, as the skiff slowly turned in the gentle current.

A River of Fish

As THE DAYS WORE ON DURING THAT EXPEDITION to substantiate Kai and John's earlier findings, the three of us became more certain that the conjectured suture zone was in the area we were studying; we were mapping it. Once we realized that, it became important to understand the relationship between the old magmatic rocks that Kalsbeek and his coworkers had found back in 1987 and the suture zone we were tromping through. The available geological maps seemed to show that the magmatic bodies did not extend north of the NSSZ. But was that simply a fortuitous geological relationship or did it mean that the shear zone sliced through the frozen magma chambers in some massive tectonic action, displacing to some unknown location the remainder of that igneous complex? The pencil gneiss in the shear zone told us that the magma, once frozen, cooled, and solidified, had experienced significant deformation there. If the sheared pencil gneisses were the only significant deformation preserved in the magmatic bodies and if the magmatic bodies were only deformed like that in the shear zone, the NSSZ would almost certainly be the major tectonic feature John and Kai had described so many years before. The question we had been pursuing had been reduced to whether the pencil gneisses were to be found throughout the region or if they were present only within the shear zone.

So, on a brisk, sunny morning, we set off to see if those sheared magmatic rocks were present in an area to the southeast of our camp. Sampling them and seeing what, exactly, they looked like would help determine how the story of that old mountain system would be told.

WE CRUISED IN THE ZODIAC ACROSS THE FJORD to a place where small coves and inlets gave good exposure of the geology. A gentle breeze brushed over the water's surface, making the trip easy but bracing. Throughout the morning, we made a number of stops, but we found no evidence of significant deformation in those old frozen magma bodies.

By late morning, the breeze had vanished and a pervasive stillness settled in. With it, the inevitable swarms of summer mosquitoes arrived, their incessant high-pitched whine setting nerves on edge. We pulled out our gloves and mosquito-netted hats and put them on. One gets used to doing fieldwork in mosquito nets and gloves; after a very short while, the netting is forgotten and the gloves come on and off easily. But when it was lunch-time, netting and gloves became a bother. We decided to escape the mosquitoes by making a dash out into the fjord with the Zodiac at full throttle, leaving the blood-suckers in our wake. We threw in our backpacks with our water bottles and lunches and John quickly cranked up the outboard. As we flew across the mirrored surface, the cloud of mosquitoes rapidly fell away and we sighed with relief, throwing the hats and gloves in our backpacks on the Zodiac floor.

Once we were out of mosquito range, John shut off the motor and the boat drifted with the slowly flooding tide, turning lazy circles in the water. The fjord was a shimmering glass surface, occasionally lapping against the side of the boat, the only sound that interrupted an otherwise absolute stillness. Small blocks of ice from the ice sheet floated past, melting their way to oblivion. We spoke few words, allowing the feel of the place in the warm sun to seep in. We slowly ate our usual lunch of bread, sardines, and cheese, washed down by a thermos of coffee.

By the time lunch was finished and we headed back to shore, a light breeze had returned. When we landed, the mosquito swarms descended, but the breeze kept them downwind, the dark clouds of aggressive, frantic insects seemingly shrieking at the fact they could not get to us. The mosquito netting was put away and the gloves came off.

The place we had put ashore was a small cobbled beach next to a long outcrop of gently sloping gneisses. The layering in the rock ran perpendicular to the coastline, which meant that we could easily walk across many different rock types, putting together fragments of a protracted history as we measured and sampled.

I took off ahead of Kai and John, who were arguing about something in the gneisses that didn't interest me. The sky was so blue, it seemed to give off a light of its own. The water, usually an intense cobalt when reflecting such a sky, was a murky pale greenish hue because of the finely pulverized rock flooding into the fjord from the meltwater gushing from the base of the ice just a couple of miles to the east.

Eventually, I rounded a small point and came to an expanse of smoothly polished stone where thin black layers in very white rock were folded intricately into distorted accordionlike forms. I walked back and forth for a short while, simply enjoying the quiet beauty of the stone while trying to make some scientific sense of it. The feeling that an anonymous potter had playfully indulged in some lyrical fantasy was irresistible.

After a few minutes, the notebook came out and, on hands and knees to more closely observe the minerals in the rock, I began writing the story that seemed to present itself. The texture of the rock pressed into the skin of my palms. It was smooth as glass in places, polished by the Ice Age glaciers that had ground at it with water and silt thousands of years earlier, but there were also small patches where the polished surface had spalled off, exposing a studded face of fractured crystals of quartz and feldspar and hornblende. I ran my hand over the contrasting textures, curious about the tactile experience of conflict between polish and edge.

The warmth was soothing. Greenland can often be briskly cold, even when the sun is out. But that warm day allowed the rocky outcrop to absorb the sun's rays and radiate back a welcome heat. I took off my backpack and jacket and rolled onto my back, feeling the warmth ooze through my shirt and onto my skin. For minutes I lay there unmoving, savoring the luscious luxury of that simple contact. After a while, turning to the right, watching in silence, the static massiveness of the ice wall at the horizon of our world captured my attention.

There was no beach there, just white rock bounding ocean. A stone's throw away, small ice blocks from the calving ice sheet lazily floated on a now-ebbing tide.

Then I noticed a huge school of herringlike fish slowly swimming by, just a few feet from the water's edge. It was startling to realize they had been there all along.

That species of fish was often out in the fjord, but usually they were alone or in small groups. They nearly always seemed dazed and lethargic, flopping from side to side, as though they lacked the energy to move fins and tail in any coordinated way. But in that school near the water's edge, they swam with purpose, flowing like a slow river toward the head of the fjord. They had gathered where the water was shallow, warm, and protected. Thousands of them, moving in a band that was many feet wide, extended from the surface to a depth hidden by the murky water. How long that living river extended was impossible tell; it stretched out of sight in both directions. I sat there mesmerized, wondering at the collective imperative that drove so many individuals to a destination they could not possibly know.

Suddenly, the river of fish exploded, scattering like a starburst in all directions, swimming away from a single point right in front of me. Frantic panic seemed to possess each and every fish. The waters churned with flailing tails and fins; if the fish had had voices, the air would have been filled with terrified screams.

Then, before I even had a chance to rise up on my elbow, a gaping mouth shot up from the depths of the opaque water. A huge Arctic sculpin was attacking the school. In a flash, the dark fish grabbed one of the stragglers and, with the herring wriggling uselessly in its five-inch jaws, slowly sank back into the murky water.

The Arctic sculpin, or ulk, is not a pretty fish—it is mainly bony head, with a spiny body and a mouth of sharp teeth. It is a near-bottom dweller, darkly colored in gray-browns and blackish splotches, an opportunistic hunter of the slow and small. It was the first time I had ever seen one.

For perhaps ten seconds, the scattered fish swam in confusion, not knowing what to do or where to go. Then, without any evident signal, the river of fish reassembled and proceeded to become what it had been moments before, an undulating pattern of life pursuing an unknown destiny, oblivious of the death that had just occurred.

A fish is a simple creature, lacking any ability to dream of success or the future; it does not imagine impassioned stories or far-off destinations. What, then, would it feel like to fear death if there was a complete absence of anticipation? What was the individual sensation that compelled unconscious migration that, in the end, served solely to assure species survival? What was the experience of following others, moving toward something unknown, something formless and indefinable, and yet irresistible? What was life in the absence of thoughtful desire or imagination?

That life-and-death drama repeated itself four more times while I sat there. Each time, the moving ribbon of animals exploded in a starburst of life, the ulk rose, killing another fish, and then sank back into the murky depths. When I left, there was still no end in sight of that river of fish.

THAT NIGHT, BACK IN THE KITCHEN TENT amid the sounds of Kai opening packets of freeze-dried soups and vegetables for dinner, the hiss of the boiling water on the Primus playing in the background, I marveled at the universality of life-and-death struggles playing out in that landscape. Bird bones, skulls of Arctic foxes, and reindeer antlers littered the tundra surface—everywhere we went they punctuated in bleached white the surface of darker shades—testament to the process that drives evolutionary change. The future is incessantly born from a surface of bones.

What we are a part of cannot be known in our engineered and manufactured world. We are the product of billions of years of unfolding change unaffected by our own intent. To truly understand what we are, and what we are part of, requires knowing that unshaped wild—that is where the bones lie.

ONCE DINNER WAS FINISHED, JOHN AND I took the plates and utensils and kitchen pots out to our favorite dish-washing rock. John washed. I usually did an inadequate job getting all the food bits off, so we agreed I would be the designated dryer. As I waited for each pot and plate, I gazed across the water, lost in my own thoughts.

After some moments, I turned toward John and saw a cloud of mosquitoes swarming behind him, riding the lightest of breezes just downwind. I picked up a soapy plate and swung it in the air at the buzzing cloud of insects. I turned the plate over and showed it to John. Thirty-seven mosquitoes were flattened on the six-inch soapy surface. John smiled, then took the plate, rinsed off the bugs, and tossed the water on the tundra. Perhaps, after all, the distinction between the ulk and me was, in the grand scheme surrounding us, not as great as I might have wished.

IMPRESSIONS III

A SMALL TUNDRA EMBANKMENT a few feet thick is to my left, a cobbled beach to my right. Next to my knee, four bones, bleached and flaking, stick out of the tundra like skeletal fingers—a vertebra, part of a rib, and two others I cannot identify. On the surface, a small tuft of white flowers sways in a light breeze, growing in a soft mat of dying, limp grasses. The bones protrude from the tundra about halfway down. The rib fragment is longer than my thumb and about as thick. From its size, it is likely that the remains are those of a reindeer.

The tundra began growing six thousand years ago, as the last ice age ended and the frozen glacial masses melted back in recession. For the bones to be so deeply buried in the tangled chaos of roots and floral carcasses, the animal must have died three or four thousand years ago.

It was at that time the first humans were settling Greenland, crossing from the islands of northeastern Canada. Before that, reindeer and musk ox freely migrated throughout that land. Would they have been fearful of the skin-clad strangers? Would they have run, or stood, curiously gazing at a carnivore they had never seen before? A landscape that had been theirs for thousands of years and the resulting inheritance of survival strategies honed by existence in a human-free world were beginning to be challenged. Looking at the bones, I wonder if I might be seeing the vestige of one of those early encounters.

Over the millennia, plants have feasted on the remnants of the reindeer, rearranging elements and compounds from animal flesh and bone into plant stalks, stamens, pistils, and leafy forms. What was not captured or useful seeped back to the salty fjord. Tidal cycles and winds circulated to the deep oceans the escaped compounds, freeing them to flow through sediments, plankton, and whales. The less soluble white and flaking bone preserved the rest.

I raise my eyes and watch blocks of ice drift on the gray surface of the fjord, water on water in a dance choreographed by moon, sun, and sea.

EMERGENCE

You quit your house and country, quit your ship, and quit your companions in the tent, saying, "I am just going outside and may be some time." The light on the far side of the blizzard lures you. You walk, and one day you enter the spread heart of silence, where lands dissolve and seas become vapor and ices sublime under unknown stars. This is the end of the Via Negativa, the lightless edge where the slopes of knowledge dwindle, and love for its own sake, lacking an object, begins.

—Annie Dillard

Tide

THE SILENCE OF WILDERNESS IS NOT JUST the absence of sound. It is a storm of voices we cannot hear because we lack the organs to hear them. In the vastness of that space rests the clatter of unfulfilled possibilities, living and not, animate and still—the echo of the dinosaurs, the mumbling of trilobites, the whoosh of pterodactyls on the wing.

As I climb out of my tent and head off to see if someone has made coffee, the calmness in the air is sobering. Walking across the short tundra span to the kitchen tent, engulfed in that magnificent silence, the frailty of our four small tents is striking. Huddled on the surface, temporary, fragile, and vulnerable, each tent is anchored by a handful of aluminum pins stuck six inches into the spongy tundra. Seeing them, it is difficult to escape the magnitude of our insignificance.

As I bend over to crawl through the tent door, the aroma brings a sense of relief—Kai has the coffee ready—the smell in the tent is luscious. John comes in a few minutes later and we start making plans for the day.

Seven miles west of camp is Tunertoq, an island that

we have not explored. It lies along what may be the northern edge of the shear zone and becomes the destination.
As we eat our usual breakfast of raw rolled oats, a bit of
powdered milk and sugar, followed by some combination of bread, crackers, cheeses, and a bit of jam, we make
plans for which headlands and bays to visit in order to find
more of that edge, and try to gauge how much time we will
need. Mapping the geometry of the shear zone's boundary
is necessary if we are going to describe the form of the
collision zone. Once we have a tentative plan, lunches are
packed, we collect hammers and compasses, GPS units,
sample bags, and the rest of our gear, then head off to the
cobbled beach where the Zodiac is secured.

John pulls the boat in and climbs aboard. Kai follows;
then I untie the line and push the Zodiac out and jump
in with wet boots. After a few pulls on the starting cord,
the outboard roars to life, briefly bellowing a plume of
blue smoke that drifts off over the water. With the motor
idling, John shifts into reverse and slowly backs us out into
the fjord. Kai and I settle in the bow, each taking a side.
Assured everything is organized, John shifts gears, slowly
turns the boat into the fjord, and opens the throttle. With
a raging scream, the engine roars to life.

As the boat gains speed, the bow drops down and spray
flies. We skim over the water, barely touching the surface.
The fjord is like glass, the swell from the Davis Strait hardly
noticeable. Sun sparkles on the drops of water flying in
our wake, a million glittering water stars shimmering in
the cool morning air. Kai and I pull down our caps, turn

up our collars, and zip our anoraks as the boat-made wind wrestles with us.

Although the thrill of discovery invades every emotional space, more profound is the sense of wonder that we are there at all. An ultimate purity of place pervades experience in that emphatic terrain of rock, water, ice, and life. Beauty becomes overpowering, cutting into the heart. An urgent restlessness insinuates itself into the moment.

How can it be that organic chemicals and a handful of trace elements have it within them to assemble a living structure that looks into a landscape and experiences wonder? What does it mean that a creature knows there is such a thing as beauty and that it resides in the deepest wilderness? It is not hard to understand the evolutionary advantage of experiencing a sense of serenity when wandering into a place of security and abundance. But here, life is harsh, survival a struggle, yet the deepest sense of awe and peace washes over me as the magnificent scenery glides by.

Tunertoq Island is twenty miles long and four wide, elongate west to east, lying on the north side of Arfersiorik Fjord. Behind it, an intricate latticework of fjords and bays stretches north and east for forty miles, eventually ending at the edge of the inland ice. There, immense rivers emerge from under the ice cap, gushing meltwater into the fjords, freshening the salty sea. When the tides are ebbing, that complex of water veins and arteries feeds into Arfersiorfik a massive amount of meltwater and captured sea. At flood tide, the flow is reversed and that fjord supplies the fluid that sustains the inland waterways.

The island is an obstruction in that complex liquid lattice, an immense and solid bottleneck situated exactly where the water tries to flow into and out of Arfersiorfik. The inland seas have only narrow passages at either end of the island through which they must flow to get in or out. Given that the tidal range can easily be twenty feet in Greenland, these passages can see vast amounts of water roaring through them when tidal flow is at its highest.

JOHN, WEARING HIS SIGNATURE BLUE BASEBALL CAP and sunglasses, sits to starboard of the outboard. Kai and I balance the weight in the rubber boat by adjusting our positions to keep us on an even keel and the bow low at high speed. Survival suits are tucked next to us in their bags. On windy days or when there is a lot of chop, we wear them, since being dumped into those icy waters can result in hypothermia and quick death. But today, with the glassy surface gloriously reflecting the morning sun, a very low swell running, and calm winds, those cumbersome suits are stowed away.

Suddenly, as though slamming into an invisible wall, the boat nearly stops in the water, skewing sharply from one side to the other. John is thrown forward, pulling the handle of the motor down, and the prop swings up out of the water, shrieking piercingly as the engine revs up. Kai and I catapult over the side and are nearly thrown into the freezing water before we grab the hand ropes on the side pontoons. We struggle to haul ourselves back in, rolling onto the floorboards with a thud. The boat jerks from side to side, swaying and bucking as though it is trying to get rid of us. With deep breaths and shock, we scramble back to our places and look back at John. My first thought is that he is playing a joke on us, but I know that makes no sense—he has a sense of humor, but risking throwing us in the water is not his style. As Kai and I try to get settled, the boat keeps up its crazy swaying, tossing us from side to side. Both the deep furrows on John's brow as he scrambles back into position next to the engine, and his intense, searching gaze to starboard make it obvious that something is wrong.

He quickly throttles back the engine and turns the bow toward the small passage we are just beginning to cross at Tunertoq's eastern end. As the boat settles down, he powers up the outboard a bit and looks at us.

"Tidal current," he says grimly.

We look in the direction of his gaze. The surface of the passage is like a roiling river. Huge boils of water bubble up in a fast-moving current, negating any evidence that there was, somewhere out there, a glassy-surfaced fjord. Our timing could not have been worse—we are there at the peak flow of an ebbing tide. The flood of water from behind the island and into the fjord is at its strongest, cutting a sharp edge between itself and the somnambulant fjord seas it struggles to insert itself into. The boundary between invader and invaded is sharp and insistent; it is that boundary we have collided with, plowing into it at full speed.

Cautiously, John angles the bow a bit more toward the west and guns the engine enough so that we can make slow headway against the current. The boat dips and slews but eventually settles into a motion that is not as jarring. Around us, the water chaotically roils.

We laugh a bit nervously, and sit up a little straighter. Anxiously, I say, "That was impressive," to which Kai replies, "It still is."

Kai and I sit attentively, hands tightly holding the side ropes, tensely aware that things are not really under control, but relieved that the small boat is stable. John skillfully works the outboard, maneuvering cautiously through the current. We look ahead, watching the turbulent water as though searching for something but not having a clue what it might be we are looking for.

Then, as though emerging from behind a curtain, a vaguely dangerous presence asserts itself. There is no doubt it has been there all the time, but the more immediate need to keep from being thrown into the water was the only thing we thought about. Now, in a more relaxed state, perception expands and we sense a threat.

The sound of loud thunder shakes us so we look to the skies, searching for thunderheads, but we see none. The sky is mainly blue, with cotton puffs of clouds lightly sprinkled about. But the sound is pervasive, reverberating all around us, and does not stop, a deep-throated, pounding rumble.

Our Zodiac is made of inflated rubber pontoons; they form the pointed bow and the sides. Two other inflated cross tubes span the inside for strengthening, but they also serve as benches. The floor is a rubberized fabric over which thin boards are wedged to give stability and rigidity. It is up through that flooring that the thunder booms.

We quickly realize that the sound must be coming from huge boulders propelled by the rushing tide, tumbling over the hard rock walls and bottom of the fjord, sculpting out of the bedrock gneisses and schists a submerged secret landscape. Minute after minute, the pounding rumble echoes up through the water, through our little boat, and into the cool air. We look at one another and at the rushing water, listen to the sounds, and hunker down a little more. John revs the engine a bit, and we make our way closer to shore. Carefully, we cruise along about a stone's throw out.

As a consequence of our own actions, we are riding along a surface that flows through a world built by forces that outstrip our ability to grasp them, exquisitely vulnerable to death. Had we been thrown out of the boat, we would have been swept away and died in minutes. The tidal roar adds an orchestral emphasis—survival here is little more than related coincidences.

Within the water upon which we ride, atoms that had once been part of the rock enclosing the sea were scraped from surfaces by pounding boulders, thereby released to float freely with the tides. In a dialogue framed by simple thermodynamics, they mingle with other atoms whose origins were wind-blown dust, interstellar particles, dissolving dead animals, and decaying plants. They converse in ways we neither comprehend nor perceive. Even so, their discussions will evolve into unities, becoming things that construct living forms or chemical sediments or simple dissolved molecules. They flow into depths, rise to the surface of the sea, and evaporate. They become snowfall on the high Himalayas, and cause the seasonal floods of the Ganges. And occasionally, they become part of us.

WE SAIL ON, THE VOICE OF THE TIDES RUMBLING in the background. We round several small points of land and cross small embayments, looking for outcrops with enough exposure to let us prowl through their history. We are moving through a world barely touched by science; only the vaguest idea exists of what might be here.

Then, fifty yards away and across a small bay, we spy bare rock running from the water's edge to an eroding cover of tundra about one hundred feet inland. Quickly, we land and head to the outcropping rock, intrigued and excited.

Exposed in that lithic fringe is a pattern so striking, our eyes wander back and forth over it, as we exclaim repeatedly how incredible it is. Bands of pink, white, gray, tan, and black, some no more than a fraction of an inch wide, some several feet thick, draw the eye along stretched-out, languid, folded forms, flowing as though the bedrock had once been as soft as butter. I feel as though I am in the presence of unencumbered, spontaneous artistry, a place where some creative genius has found its rhythm and manically painted from inspired passions, using fluid rock as its medium. Every step we take is a halting one, each new square foot possessing a different form or pattern of colors. We crawl on hands and knees, trying to grasp the significance and history of that place. From a scientific point of view, it is a treasure. From an aesthetic point of view, it is a masterpiece. Our quantitative world has seamlessly become enmeshed with an ethereal realm, dissolving into a Dalíesque fluidity. What we are doing no longer has boundaries; everything the mind can embrace is present here.

WE DID NOT KNOW AT THE TIME that those are the oldest rocks in the region, remnants of some of the most ancient continents on Earth. It took many months of work back in our laboratories to discover that they were formed more than 3 billion, 300 million years ago. They preserved evidence of the existence of an ocean basin billions of years old, when life was only single-celled and free-floating and what little land existed was adrift with blown sand and utterly barren. It was an ocean vastly older than the one associated with the building of the mountains we had come to study. Black layers had once been molten rock, injected into the sediments of those old seas, probably long after the water had been squeezed from them and their crystalline form changed. Deeply buried, heated, and compressed, the entire sequence was later folded and refolded, deformed and intruded during some unknown mountain-building events spanning hundreds of millions of years. Eventually, sometime in the last few tens of millions of years, they had made it back to the surface, shoreline to a new ocean, supporting our boots while waiting for another transformation. It was, in fact, the northern limit of the zone we were looking for. It was the very edge of one of the continents involved in the collision.

Since making that discovery, John, Kai, and I have visited that headland several times. We are trained observers, looking with a critical eye for facts and hints that will fill in details about the big picture and the small. We want to know the linear history embodied in that complex of patterns, colors, and textures. We record and sample. We make repeated measurements. We argue and deduce. And yet, no matter how carefully we take notes, measure orientations, describe patterns, mineralogy, and textures, each visit is a new revealing. We look at an outcrop or pattern for the third or fourth time and see things we have not noted before.

All landscapes shape a future terrain. In our moment, riding the back of those implacable forces, we are entangled in that process no less than the boulders pounding the walls of that imperious tidal channel.

Clockwork Pebbles

Days and miles pass. The three of us collect fragments of information—measurements of mineral orientations, the strike of planar features, the mineralogy of layered rocks—take samples and notes, all in an attempt to augment what little is known. The naked eye, hand lens, and compass, are feeble tools, and we will need results from laboratory analyses before elements of the story can be pieced together. Even so, what we see provides first impressions, a few facts, and the beginnings of insight. And always, in the evening, we sit and talk, the conversation weaving through the complexities and joys of our private lives and the experience of the science we are pursuing.

On this particular day, there is a vague aura of satisfaction. The shear zone is not a "straight belt," but a zone of massive movement, one of the defining elements of convergent, ancient continents.

I climb out of the cook tent and head toward the little beach that fringes our tundra-blanketed bench. The scramble down the small cliff that bounds our home is easy, the walk to the water's edge spent mulling over the conversation I'm leaving behind.

The beach is a short stretch of cobbles and pebbles and not much sand. A little ridge of rock about ten feet long runs parallel to the shore, six feet out in the water, off to my right. The tide is coming in and starting to drown the rock ridge. Ragged, choppy little waves wash up on the beach, except at that stony barrier where they spend themselves in a brief frenzy, then wash around it, carrying small pebbles with them.

I walk over to the protected backwater behind the barrier and stand by the water's edge, looking out across the fjord. Clouds scud by overhead, casting a gray gloom over everything. The distant shore is featureless in the dim evening light, just a dark presence on the other side of the water. Lost in thought, I stand there long enough to let the rising tide and little waves make a surprise advance on my boots, soaking them in a flurry of white water and foam. I quickly step back with a crunch of grinding stones, pushing pebbles into mounds and depressions that accidentally mark where I stood.

That new topography is an assault on the smoothly sloping surface that the waves prefer. Within seconds, the advancing salt water attacks the raised lip of jumbled pebbles, collapsing them back into the space where I had been. As the waves work away at my inadvertent engineering, the beach slowly recovers its original form, returning to some state of quasi-equilibrium. Within minutes, there is little evidence of any human intrusion.

It is bitingly cold. Standing there isn't comfortable, but

something about the feel of the place holds me. I pull up the collar on my parka and glance down.

The stones that shingle the beach are fragments of gneiss and schist, weathered from the outcrops we have been studying, eroded and ground down to flattened, smooth oblongs. Most are remarkable only for their dark, gray, unidentifiable plainness.

The pebbles extend into the fjord and beyond the tidal zone. The water, crystal clear here, makes it possible to look into the depths, where the light vanishes and the stones become progressively less distinct. There is no boundary where the pattern of pebbles ends, only growing dimness, a dissolving.

There is one small gray flat oval of a pebble that I have been watching. Lying among the others, slightly tipped up, it has a thin edge protruding above the rest. A small wave breaks on the shore and washes across the beach, immersing it. As the wave spends itself and rushes back into the fjord with a light hiss, the little pebble flips over in the brief chaos of turbulence and foam. One wave, one pebble, and the metronome of process registers one more click.

THE DAY WE DISCOVERED THE STONE that smelled of singed hair, we saw something I had not appreciated at the time, but the tumbling pebble resurrected the memory.

After collecting that stone and cruising back toward camp in the late afternoon, a brilliant flash of light from the shore we sailed along caught our eyes. It was an odd reflection six feet above the high-water line, several hundred yards away.

John swung the boat around and slowly retraced our course as we looked for the flash to repeat. When it did, we fixed the location and sailed toward it. There was no beach there, only huge angular boulders that had tumbled from the steep thirty-foot wall of rock at the fjord's edge. As we slowly made our way along the coast to the west, John idled the outboard, eventually finding a small pocket of sand to land on.

As he beached the Zodiac, John reminded us that the tide was rising and we had little time.

We grabbed hammers and jumped out, tying the boat as securely as possible to several boulders. The outcrop was a good distance away, across a jumble of talus we tried to scramble over quickly. As we carefully picked our way along, we kept looking back at the Zodiac to make sure it wasn't floating off.

The outcrop was a dull, dark olive green. The reflection came from a perfectly flat surface, polished almost to the sheen of glass, more than a foot long and eight inches wide. As we moved our heads back and forth, changing perspective, we could see that the reflection wasn't a single sun reflection, but was composed, instead, of several rippled parallel bands. We realized it was a single immense crystal, the reflecting face a cleavage surface that contained bands with slightly different crystal structures called "twins." Surrounding the crystal was a white rim less than an inch thick. As we looked more carefully, it quickly became clear there were hundreds of these immense crystals, each rimmed by a white rind, each stacked like a brick in a massive pile, one upon another. Stunned and excited, we realized this was an accumulation of giant orthopyroxene crystals, something that had long been postulated to exist but had never before been seen.

When continents first form, they evolve mainly from diverse melts that rise from the mantle. Some melts are able to penetrate any crust that exists and flow out as lava onto the growing continental surface. But other melts, as they rise from below and encounter the base of the crust, cannot penetrate because they are either too viscous or too dense. A particular kind of melt that was believed to commonly stall at the continental base, but was universally involved in the development of continents, is called anorthosite. Trapped at the bottom of the developing continents, the liquid slowly cools over many thousands or millions of years. In this gradual chilling process, crystals form as the magma progressively solidifies, the newly formed minerals incrementally growing and settling to the floor of the magma chamber, piling upon one another. It had been postulated that giant orthopyroxene cumulates must form during this process—giant orthpyroxene crystals had been found in anorthosites around the world—although the preserved giant orthpyroxene cumulates expected to have formed at the very bottom of the cooling magma chamber were unknown. Yet we had now found such an instance, the thin white rims being

preserved anorthosite melt trapped as the huge orthopyroxenes settled onto one another.

We tried following the cumulate to get an idea of its
shape, but within a few feet it ended against an intensely
sheared band of rock many feet wide. We went in the
other direction and discovered the same thing. Examining
in detail the sheared material, we realized it was nothing
but the finely ground-up remnants of the huge crystals.
The cumulate giant orthopyroxenes, which probably
extended for miles when they originally formed, had been
abraded down to a small lozenge a few tens of feet across.
We quickly made measurements, collected a few samples,
then raced back to the Zodiac. The tide was lifting it and
the line would not hold much longer.

We returned to the site twice, making observations
and collecting samples sufficient to allow the significance
of that small outcrop to become clear. Eventually, through
many hours of laboratory work, we were able to show that
the giant crystals were formed more than twenty miles
down in a magma chamber at the base of an ancient,
evolving continent over 2.8 billion years ago. They and
the crystallized magma they had settled out of had been
recycled and transformed through the grinding continental collision we were studying, becoming constituents of
the new landmass.

The simple transfer of momentum, the small bit of
heat lost to the shiver of atoms, the dynamics of contrasting masses flowing past each other: The physical reality
of mathematical equations is made manifest by pebbles

tumbling with the tide and a small sliver of sheared cumulates. The richness inherent in nature's simple statements fills me with awe.

WHEN KAI, JOHN, AND I RETURN to our laboratories, we will describe much of what we have seen through equations that honor the observations and data we have collected. By doing so, we will attempt to objectively communicate the details and subtlety of the history in those rocks.

But the quantified reality we will convey will be more than an analytical result. We will use equations to calculate, from data collected in mass spectrometers, the ages of the samples we have collected. Those equations, derived from atomic physics a century ago, become time machines that focus imagination, opening doors that let us see the pace at which our planet's surface evolves. Other mathematical formulae allow us to compute the chemical composition of minerals, giving us insight into the chemistry of the oceans and atmosphere bathing Earth billions of years ago, and providing a glimpse of the path which led from naked stone to the human mind.

Those same equations have shown that the universe is drenched in light that spans a hundred orders of magnitude of energy. Animal vision is constrained by the ability of organic molecules to absorb and respond to only the tiniest fraction of that spectrum. What we see isn't even a ghost of an outline of what is out there.

I AM NOT WHO I WAS WHEN I GOT OFF THAT PLANE at Kangerlussuaq. Certainties I held to be immutable—what the world was, what constituted reality and knowledge— are evolving as we live here.

Separation from the clutter of culture removes the incessant challenges of having to judge, act, and react to bombarding opinions and information. There is no need to struggle with the effort to make sense of the right and wrong of anything, for in this aggressively wild space, there is no judgment, simply the act of being.

Walking back to the cook tent to talk more with John and Kai, I am once again struck by the rugged frailty of this place. The small bluff at the edge of the fjord just a short distance from our tents is actively eroding, the small rampart of accumulating boulders at the base of the cliff the only remnants of now-vanished landscape. The places we walk in camp are becoming worn paths. A small ice field across the fjord from us has noticeably changed shape and shrunk in the weeks while we have lived here. And so it will be—any evidence of our presence in that wilderness will be erased within months of our departure, just as the small waves eliminated the punctuation of my boot prints.

Ice

Arfersiorfik Fjord runs from the Davis Strait to the ice front, a distance of nearly a hundred miles. Arfersiorfik means "where the whales are," or something similar, depending upon whom you talk to. One of the Greenlanders who took us into the field one year said the name came from the fact that often during the winter, the mouth of the fjord stays ice-free, giving the whales a place to breathe.

Reaching the eastern end of the fjord can be difficult because ice calving from the face can clog the water. But this year, temperatures are warm and summer has come early. Since the eastern end of the fjord has long been a place we have wanted to explore, we decide to make the trip. Years earlier, geologists had been there, but what was mapped had been done as a quick reconnaissance. We have no details. The maps that we use imply we would find the remnant magma chambers that were the first indication of an old volcanic mountain system there. It is an obvious place for us to visit.

Because the day is going to be long, with many stops for mapping and sampling, we have an early breakfast and quickly launch the boat. The morning is sun-drenched and calm, the water surface softly undulating with a rhythmic, shallow swell.

Cruising along the shore, we repeatedly make landings to investigate and record. Some of the stops were planned long ago, based on gaps in our data and curiosity about what intervenes between two known locations. But many stops are spontaneous, demanded by some odd or unexpected configuration of color and pattern in an outcrop. As always, each stop exposes something new, providing small insights that embellish the geological story by infinitesimal increments. We find a place where a major thrust fault comes to the water's edge, marking a massive zone that must have seen thousands of major earthquakes over several million years of grinding and slipping. At another site, brilliant blue tourmalines decorate thick white lenses of once-molten rock, attesting to the presence of boron and other elements from ancient ocean water trapped in crystals formed during the collision of tectonic plates. We smugly revel in our good scientific fortunes. Each new find, so far, remains consistent with the story of intense, long-grinding deformation.

As we sail along in that quiet serenity, basking in the pleasure of small discoveries, the rolling hills and modest cliffs take on an air of fantasy, as though we are gliding by a pastoral coastline where, around the next bend, a white gabled country inn might materialize. It feels as though we are moving through a place in which even the smallest pebble or blade of grass is encased in some spellbound reality.

As we sail around a point and look down the fjord, we are slammed back to the hard-core realty of our geological pursuits. A few miles ahead of us, a sheer rock face several hundred feet high blazes in shades of off-white and pink, a stark and startling contrast to what we have been seeing. The very top of the cliff is the dark gray country rock, native to that area, that we have become intimate with, but the rest of the face is shot through with much lighter rock engulfing the gray host in a patchwork of angular geometrical threads, fingers, and thick veins. Dark gray blocks hundreds of feet long and many tens of feet wide are trapped in the whitish pink wall, examples of classic, textbook xenoliths. We have stumbled upon the eroded top of a large intrusion of granite, something rarely observed with such complete exposure. We are looking at the upper reach of a large magma chamber.

Each of us has seen idealized diagrams in which a magma body stopes its way upward, filling in the space that is left behind as blocks of the roof of the enclosing host rock fall away and settle on the floor of the magma chamber. But this example is on a staggeringly dramatic scale. In all our careers, none of us has seen anything like this.

John speeds up, and in minutes we land the Zodiac at the western edge of the body. The granite and the suspended geometrical blocks form exquisitely beautiful patterns—the pink intrusion is riddled with tiny, perfect pink garnets; the suspended blocks are encased in intensely black rims; light tan and black micas glisten in the granite; and veins of white and black minerals crosscut everything.

What we are walking over is just the very upper portion of a chamber of magma that had been rising slowly through the crust just after the collision of continents. The magma had formed from rocks that had been pushed deep into the earth and heated beyond their melting point. The melt that formed had collected into a single body that slowly rose up through its host rock. As it rose, it lost heat to the cooler rocks it was passing through, eventually freezing in place. After nearly 2,000 million years of uplift and erosion, it is now exposed to the sun, providing a solid bench for our boots.

As LUNCHTIME ARRIVES, WE PACK UP our samples and head east, hoping to find a place where we can explore the actual ice front, for a little while. But about half a mile from that massive ice wall, the opaque fjord waters become thick with silt. Conditions like these can sometimes hide water-saturated mud shoals under a veil of water a few inches deep. Running into such a thing would ground us in the middle of the fjord, with a real possibility that we might not be able to get the Zodiac out. As a precaution, John turns sharply toward the northern shore and lands us there.

We find a small grassy ledge and settle in to eat lunch and watch the ice. Despite the distance, the view is spectacular. A buttress of broken ice blocks lies at the base of the ice face, testimony to a long history of icefalls and avalanches. From that jumbled chaos, small ice blocks float away at high tide, filling the water in front of us with an infinite variety of shapes lazily flowing with the current. Gulls are all around, bobbing in the frigid water. Every now and then, one of them takes flight and lands on an ice block and casually sails by us out into the fjord, then abandons its carrier and flies back for another ride. Many gulls do that repeatedly—whether it is some effort to entice us to feed them, which we don't, or if they are just having fun, is never clear.

I have always wondered what it would be like to ride an iceberg—what the surface is like, how buoyant it might be, what the touch of one would feel like. I mention this to Kai and John and we debate what to do. Finally, it's decided we will take a few minutes to try getting me onto one of the frozen blocks floating by.

We finish lunch and pack up.

But before we leave, John reaches into his backpack and pulls out his camera. Handing it to me, he asks in a slightly sheepish tone if I will take a picture of him in front of the ice. He walks over to the edge of the little bench we are on. In the background, the massive front of the Greenland ice sheet glares in brilliant white in the noontime sun. Standing tall, with his head back a bit, John stuffs his hands in his pants pockets, turns slightly toward the ice, and says, "Now."

We each take turns posing.

Then we load up the boat and sail out into the fjord toward one of the ice blocks. It is about ten feet long and five wide. At the water's edge, the melting of the small berg has carved a scalloped, rimming incision. Above that, a narrow ledge surrounds the block that bounds sculpted, delicately laced ice ridges, fingers, and hummocks. The surface is like an intentionally carved sculpture garden of abstract forms that are slowly, invisibly melting.

I ask John to get next to it so that I can see if I can get on it. Gingerly, he navigates the Zodiac to the edge and tries to keep it there.

The surface sparkles in the sun, carpeted with a network of tiny ice crystals that form an interlocking, fragile transparency. With care, I slide onto the Zodiac pontoon and put a foot on the tiny iceberg. The crystal network shatters under my boot and immediately the ice begins to roll, bumping into the boat. It is unexpectedly delicately balanced. Since we have no idea what its underwater form is like, and whether or not it can tip us if it rolls, we quickly back away and let it float off.

We spend the rest of the day on the south side of the fjord, then start back toward camp. A modest wind has come up, blowing against us, making for a slow and choppy journey.

ICE HAS NO PERMANENCE ON THIS PLANET, but its metamorphosis from snow to ice fields to massive calving ice walls is more than a shift in form. It alters light, shapes sounds using its own voice, and responds to touch. It is a separate world of experiences, one that is rich and deep. This I learned years ago, from a different setting, far from the water's edge, at a place in Greenland where the ice sheet ended on land, miles from any fjord—a place where the experience of ice was more intimate. There, the distinction between ice and rock was, to some degree, arbitrary, and the experience of it a revelation.

It was a place miles east of Kangerlussuaq, a site I and a few other researchers had traveled to in an old army truck along a vague and winding dirt road. We drove to a small hill a few minutes' walk from the edge of the ice sheet. The tundra biome carpeted the ground for some thirty feet in front of us, then abruptly ended where the advancing ice, a few years before, had scraped off the overlying soil and plants and then retreated, exposing a glistening rock surface that had been repeatedly sanded and polished by the ice for millennia.

We were at the southern end of a lobe in the ice cap. To our right, the ice was bounded by a moraine of rocks and dirt over fifty feet high that ran for many miles along the ice front. The moraine had been pushed up as the ice migrated over the land. But as climate change took hold and warmed the air, the ice was melting back and was barely in contact with its rimming moraine.

Directly in front of us was a huge ice amphitheater partially filled by jumbled blocks that had fallen from the ice face. To our left, hundreds of yards away, the amphitheater ended in another massive wall, in which there was an enormous ice cave. The cave extended back into the ice hundreds of feet. It was impossible to tell exactly how far back the ice cave reached, because its depths were blackened in deep shadow—it could have been a quarter of a mile or more. Inside the cave was a waterfall at least forty feet high that fed a river cascading down ice blocks covering the floor. The river rushed from the cave and flowed along the base of the wall in front us, a persistent liquid boundary between rock and ice.

Low rumbles, snaps and pops, and repeated booms came from the ice front. I walked closer to it, trying to figure out what was making the noise. I had imagined the place would be a silent whiteness, but the wall itself was a cacophony of sounds that pounded through a startlingly complex pattern in the ice of pale blues and ribbons of brown that laced through all shades of white.

The ice wall was water that had fallen from the sky thousands of years ago and hundreds of miles away to the east. After having been buried and compressed, it had recrystallized and sunk to near the base of the ice sheet, where it scraped up rock fragments from the bedrock and pulverized them to fine dust. Now, after slowly migrating a few inches per year, the frozen ice was exposed in the cliff in front of me, sunlight once again shining on the molecules of water that would soon be freed to flow in rivers to the sea and repeat the cycle. The booming, snapping, and popping was the voice of that frozen water as it scraped over the land, internally cracking into crevasses and fissures, preparing to be released.

After a while, we began to walk along the amphitheater. The maze of ice blocks was a jumbled chaos that would have been impossible to walk through. Some blocks were the size of fists, some the size of houses, and all of them were sharply angled and lying precariously in illogical arrangements. I turned to one of the researchers with me and started to say that I would love to see an icefall, when, at the back of the amphitheater, a loud cracking sound reverberated across the icescape.

Slowly, almost imperceptibly even, a huge section of the wall began to move. At first, the face simply seemed to shift slightly, with a few small pieces leaping in free fall off the wall. Time seemed to slow down, as it sometimes does when one is startled or threatened.

For what seemed like many seconds, but could not have been, I watched as the wall cracked and fractured and crumbled across the entire amphitheater's face, cascading down in an accelerating free fall. With a tremendous explosion and roar, the ice smashed into the jumble of blocks at its base. Ice flew everywhere. Some of it bounced off blocks in the remnant pile of earlier collapses; some smashed into fragments and ricocheted off the ice front. A few pieces the size of baseballs flew out toward us, landing in the river and shattering on the polished bedrock around us. Then, within seconds, the drama was over and the roar from the fall died away. Ice dust, drifting as though it were fog carried in an early-evening breeze, drifted off in the air, and the scene, slightly rearranged, returned to the stillness it had been.

Scattered all around were sparkling jewels of shattered ice. I walked over to where there was a collection of them and picked one up. It was a single crystal of frozen water the size of a Ping-Pong ball, irregularly faceted by gently curved surfaces into a magical gem. It was perfectly clear, with fine trains of microscopic bubbles laced through it. The thinnest film of liquid water covered the smooth, irregular facets. I held the transparent nugget up toward the ice wall and looked through it like a lens, impressed by its brilliant clarity.

I put it in the palm of my hand and looked at it from all angles. The smoothness of the liquid surface begged to be tasted. I put it in my mouth.

Beyond the first impression of cold, the taste was almost exactly what the clarity of the ice implied it might be—clean, refreshing, soft. A feeling of calmness came with it. And then, startled, I realized there was an impression of smell. I breathed in and was immediately taken by a sensation of open sky, clean air, and earth. I took the ice out of my mouth and picked up another crystal. I brought it to my nose and smelled it. The experience was of something subtle but persistent, something that was fundamental—a thing that was nothing but its own essence. Flint and stones came to mind, along with graveled river edges and a very faint mustiness. It was a smell that brought out feelings deeply rooted in some old experiences of watery, lithic places. I kept inhaling, trying to capture the impressions, but they flitted away as quickly as they materialized.

The sense of smell is deeply imbedded in the wiring of the brain. The olfactory organs carry messages to the olfactory bulb, from whence information is transmitted that becomes part of our cognitive and subconscious experience. Although species-specific, the wiring of this circuitry is schematically similar for most animals. The wiring for smell seems to be something that evolution perfected early on—it has guided creatures for hundreds of millions of years. Could it be that part of that evolutionary schooling included learned lessons—that certain smells implied certain possibilities, good and bad, that would influence behavior? Was it possible that such sensitivities would be selected for, carried into future generations as a benefit for survival? Could one such lesson for humanity have been the smell of ice and its implications? Perhaps there is knowledge contained in that smell, knowledge of the danger of icefalls, of the possibility of woolly mammoths and food, of fish and berries, of marshy land and the annoyance of mosquitoes.

I imagined an ice wall bounding a Stone Age world. An Ice Age hunter would have tracked animals, searching for food in a primordial place. With others of his kind, he would have walked through places much like where I stood, and read the ice and the land, perceiving the dangers, and sensing where barren-ground caribou and mammoth and musk ox and fox might be. They would have found the right places to spend the night, protected from the wind and wet and cold, their tolerance for discomfort beyond anything I would ever know. They would have gathered plants on their journeys, and collected the right stones for shaping. They would have talked in some long-lost language.

It was a time of an unembellished Earth, when the wild character of land and life existed as an unconstrained stage across which humanity wandered and time was immaterial.

Seal

THE PURSUIT OF SCIENCE IS AN EXCAVATION. Insights gained through research expose unexpected layers of past histories richer in content than what we could possibly imagine.

After our third expedition, it was unequivocal that the shear zone was a scar, slashed across the northern edge of the collision terrain as a last act, a tectonic finale in a mountain-building drama. That scar was what the early researchers had claimed it was—a zone of major movement. Kai and John's work was correct and the region reverted to the term they had used for it years before—*shear zone* replaced the *straight belt* moniker on later editions of geological maps and in publications.

But buried in the crystalline record, frozen in the minerals of a few rocks from small, scattered localities, was evidence that these rocks had descended into earth at least a hundred miles before the collision of continents began. That part of the story had been completely missed. Uncertainty had changed in form but not magnitude—new questions now had to be addressed.

One of those questions was the significance of the rocks that had been so deeply buried. Only a handful of places around the world had histories of so-called ultrahigh-pressure metamorphism—metamorphism under conditions where pressures were more than 400,000 pounds per square inch, a state that is achieved in the earth only at depths beyond sixty miles. The evidence in all those other locations came from ancient subduction zones. In every instance, those subduction zones marked locations where continents had collided and were thus consistent with the history that was suggested as a possibility in our study area in Greenland. But none of those other sites was older than 900 million years. Explanations varied for why all those other locations were so young compared to Earth's four-and-a-half-billion-year age. Some believed that the inherent instability at the Earth's surface of minerals formed at such high pressures would result in their slow but inevitable retrogression to other minerals, ones that were more stable at low pressure conditions. By that reasoning, it was concluded that about 900 million years was the maximum amount of time such unstable minerals could persist. Another explanation was that plate tectonics, with the ocean floor spreading and the subduction zones we see today, did not become completely established until about that time—earlier plate tectonics presumably were expressed through some other, and as yet unidentified, mechanisms, possibly involving much shallower convergence zones and no deep subduction. Whatever the explanation, we were faced with the challenge of explaining the uniquely old age of the ultrahigh-pressure rocks we had discovered. Since it was now clear that our samples had persisted for twice as long as their younger counterparts, they were either the products of some very unusual preservation mechanism or rare evidence of much older plate tectonics that has yet to be found in other settings. Given the unusual nature of the rocks we were studying, the answer almost certainly lay in a combination of both ideas.

Also clear was another puzzle. In a library of hundreds of samples that had been collected and studied by several different researchers over four decades, we had found two that preserved evidence of ultrahigh-pressure conditions. One reason why that evidence had not been previously recognized was that the mineral characteristics and compositions that indicate such conditions were not understood until recently. But the fact that we found a record of such conditions in only two samples out of the hundreds we reexamined raised another question. Did this mean that evidence of such extremes had been nearly completely obliterated by later events, like the late shearing, leaving a very small fraction of rocks that kept their history intact? Or was it evidence that the entire region was actually a tectonic jumble in which rocks from vastly different locations and histories had been forcefully interleaved with one another?

OUR FOURTH EXPEDITION IS AN EFFORT to better understand the new questions. We have decided we will spend a few weeks moving from campsite to campsite so that we can visit key localities that are scattered over a thousand square miles. Carsten, the owner of a small cabin cruiser in Aasiaat, has contracted with us to be our logistical support, moving us when we need to relocate, taking care of his boat at our campsites in the meantime.

One of the sites we've decided to visit is a place where John worked years before. It will be an eight-mile traverse; we will be hiking along exposures of marble that had been tectonically intermingled with very old gneisses. John had mapped this area while working on his doctorate, before plate tectonics models that embraced colliding continents were fully accepted. Back then, the conceptual paradigm was called "geosynclinal theory," in which it was envisioned that huge basins hundreds of miles wide and thousands of miles long were scattered across the globe. Although they did not migrate across the face of the Earth as tectonic plates do, it was believed the basins would slowly subside, growing deeper and deeper with time. As they did so, they filled with sediments. Eventually, through unknown mechanisms, they reached an unstable state and compressed, with massive mountain systems growing out of them. Since the data collected at the time were described using terms useful for the geosynclinal theory but incomplete for plate tectonic concepts, we wanted to look at the area in more detail to see how it might fit into the new picture of things.

Morning emerges gray and still. The cruise along the coast is calm and easy. We head toward the mouth of a small fjord, where we can land and start the traverse. The terrain will be uneven but not rugged—a hike that we can easily do in a day.

By the time we reach the fjord mouth, the tide is slack. The short trip from the boat to shore, using the small skiff that is tied to the stern, will be easy. Backpacks and hammers, food and water are tossed in. Then, just as we are getting ready to launch the skiff, the head of a seal pops up a few hundred yards off the starboard rail and somewhat astern. It watches us curiously, head high out of the water, keeping its distance. Carsten sees it immediately and becomes quite excited. He knows this could be dinner, a pelt and dried meat for his family.

He jumps off the skiff, runs to the cabin, and grabs a small-bore rifle that is mounted above the port door. He checks the chamber and loads the magazine, then runs back to the skiff and quickly takes us ashore. Although he attentively guides the skiff to our chosen landing spot, every few seconds he looks back over his shoulder, keeping an eye on the seal. Once we land, he races back to the boat and heads out after the seal, the rifle laid across the top of the instrument panel in front of the wheel. If all goes according to plan, we will meet up with him around suppertime, back at camp.

We immediately spot the marble outcrop we are looking for directly across the stony beach we have landed on. It is a medium gray color, about six feet thick, and sandwiched between brownish black gneisses. We hike along it, impressed by the intricately folded structures and stretched inclusions that embellish it—undeniable testament to extreme shearing. It is consistent with the strain rocks would go through if they were caught between massive continents grinding together in a collision zone— another stake in the ground.

As we walk and talk, an endless array of small meadows, tiny ponds, and new plants present themselves, offering botanical delights we had not expected. At one site, a thick blanket of dark green and tan moss is folded at the base of a six foot high outcrop. I am baffled by it, never having seen moss grow in such luxuriant folded forms. Then I realize the moss could not have grown that way. Instead, it must have grown on the outcrop surface, softly covering the rock face in a photosynthetic blanket. For decades, if not centuries, it grew undisturbed, but eventually it reached such a thickness that the weight of the blanket was more than the fragile connection between rock surface and plants could support. The mass of moss ultimately collapsed into a vegetative crumpled blanket resting at the foot of the now-barren rock face. The folded plant carpet is surrounded by bright yellow, finger-thick, feathered stalks of some unknown upright fungus. Were I a mycologist, I would be in heaven. Being simply a geologist, I pass through in a state of wondering bewilderment.

Suddenly, from the distance comes the unmistakable crack and ring of rifle shots—the short, sharp reports almost being a counterpoint to our hammer blows on the outcrops. They accompany us for the next few hours.

By late afternoon, we reach a rise that stands several

hundred feet above and more than a mile away from the cove in which we had set up camp. We can see that the boat is anchored a short distance from shore. We wonder if Carsten has the seal, but we can't tell from that distance.

After another twenty minutes, we make it to the cove and camp. Carsten is on a stone shoulder that slopes into the water. The seal is laid out on the stone and he is carefully skinning it. His movements are precise and skilled. He makes sure the skin is properly cleaned, uncut, and that the meat is washed. Then he loads the skin and the cleaned and butchered carcass into the small skiff and heads to his cruiser. A short while later, he returns for us so we can have dinner on the boat.

Carsten explains to Kai in Danish that he is going to prepare a local delicacy that we likely will not want. After a somewhat hesitant translation, we finally understand that he is preparing cleaned and boiled entrails with a few other ingredients. The smell, he is certain, will bother us. He excuses himself as we prepare our dinner. Kai takes over the responsibilities for cooking our meal in the galley while the skipper prepares his own meal outside on the fantail. Life lived in Greenland is integrated with the life of the sea; it is balanced and nuanced, and nothing is taken for granted.

I am reminded of an experience that startled me on the first expedition I went on. We were about to leave from Sisimiut on a small fishing trawler. The day was chill; everyone was dressed in anoraks and parkas, knitted caps and gloves. We were loading supplies on the boat, handing them over the rail from the dock to our companions, who would secure them to bulkheads. Food boxes were stowed below. As I handed a stuffed backpack to a deckhand, I glanced across the water to a dock adjoining ours. Two men were repairing fishing nets, their ungloved hands nimbly knotting the line to eliminate holes. As I watched, one of the men turned around to the roof of the small cabin next to him and picked up a knife that lay next to the body of a small ringed seal. With the slightest motions, he sliced off a wedge of blubber and ate it, then returned to his work. It was his snack before heading out to sea. That seal would last for the days the two men would be fishing.

ALONE AND DOWNWIND, CARSTEN EATS his meal outside. As we eat, we occasionally look over our shoulders at him, impressed by his gusto. Then, unexpectedly, he shows up at the door of the little galley with a plate of meat. Asking us if we would like to try a taste of the seal, he passes it around, and we each take a small slice of meat.

What is on the plate looks like tough beef, very dense and with a distinct grain. There is little fat in it. An odd smell, slightly sweet and cloying, while at the same time gamy, emanates from it. I take a bite, expecting something close to what I vaguely remembered of reindeer that I had tasted years before. Although the texture is much like that of tough beef, and there is a kind of beeflike character to it, I am startled by the overwhelming and completely unexpected flavor of fish.

To experience a place is to find nourishment there. A seal knows the subtlety of fish movement, of fish habits and patterns. Its brain is wired to hunt fish, knowing where fish are likely to be, what evasions they will make as they flee, what endurance they must have to get away. That inherited knowledge is the benefit of selected lessons from millions of years of successful and failed hunting. Inevitably, a seal's experience of place, and how it moves within it, bares the imprint of what it seeks for food and what it eats; it lives life, in part, with the perspective of a fish.

If I were to taste my own muscle tissue, what would I think? What would I learn about my experience of the world, what I seek and how I live, from what that taste would conjure? It is a certainty, as it is with the seals, that we have inherited ways of seeing, ways of looking at a landscape or clear water or the sky, that descend from evolutionary knowledge that relates to survival. We are the sum total of that inheritance and the expression of those lessons.

Living in the wild heart of untouched spaces brings to life taste as a forgotten language and vocabulary that encompass the elements of place. Through that language, the history of where and how a life was lived can be written. The vocabulary of that language acknowledges the fauna and flora of place, the landforms and water features, the change of light with seasons.

Belonging

AFTER MORE THAN FOUR WEEKS, our time in the field is ending. The conflict about interpretations of the history of the region has been resolved, but new complexities need to be explored, the hints of deeper histories considered. We are eager to get on with the next phase of work, compiling our measurements and observations into a coherent chronicle, and studying and analyzing the suite of samples we have collected. There is, too, the excited anticipation of being back with families and friends, returning to the modern world and the conveniences it provides. Shortly, a helicopter will be here to fly us to Kangerlussuaq.

Kai has marked out a landing site with a big X on the ground, using white cloth strips brought along for that purpose. The site is a short distance from where our tents stood—a small bench that steps out from the rock ridge I had climbed our first night in the field together. There is just enough space so the helicopter can land without smashing its rotor blades against the rock wall. This landscape is not hospitable to modern technology. The morning, gray and chill, with a light breeze blowing in from across the water, makes for a bitingly cold good-bye.

The day before, we boxed up unused supplies and

the equipment we would be flying out with. Hundreds of rock samples were wrapped in newspaper, labeled with identification numbers and the coordinates of where they came from, and packed into wooden crates. Arrangements had been made for the blue-hulled trawler that brought us into the field to pick them up later and take them back to Aasiaat for shipment to Denmark. We double-checked entries in the docket book to assure that accurate latitudes and longitudes were recorded and that the descriptions of the samples were consistent with our notes. When that was done, we collected all the trash and burned it on the beach at low tide.

The sample boxes now stand as the sole remnant constructs attesting to our presence here. We took down our tents early this morning.

Minutes before it is supposed to arrive, we hear the rapid *thup-thup-thup* of distant helicopter blades beating the air. The sound comes from across the water, miles to the south, echoing off the massive walls enclosing the fjord. We strain our eyes to see the copter, but we see nothing.

A few days before, a group of three Greenlandic families, nearly the only people we have seen here, set up tents at the headland by the stream where we collected water and where we bathed. They had come to hunt reindeer. Our only contact with them was the day after they arrived.

We had come back from the field late in the afternoon and found four of the kids standing on the bluff above where we stored our fuel and supplies. The kids watched as we beached the Zodiac, tied it up, and unloaded rocks and gear. We waved, but they kept their hands stuffed in their anoraks. Our stash of gear included extra life vests. As we organized our things and stacked them among the supplies, we discovered that one of the life vests we had left behind had been inflated. Curious young minds had found it all but irresistible—the small bright red plastic knob that had to be pulled to inflate the vest was just too tempting not to tug. I was sorry to have missed that moment.

They hung around for nearly an hour, trying to decide whether to come to our camp and see who we were, while we cleaned up and started preparing dinner. But they never did. I regretted not having gone over to them to introduce myself.

EVENTUALLY, WE SPOT THE HELICOPTER. It is heading directly at us, like a glistening red-and-white rocket targeting our insignificant outpost for obliteration. Within a few moments, it swoops over us, makes a steep diving turn, and settles onto the spot Kai marked. I glance over toward the small group of Greenlanders, wondering what they are thinking. They are all out, standing by their tents, watching.

Within a couple of minutes, our gear is loaded and we climb into the chopper, buckle up, put on headsets, and lift off.

As we ascend from camp, for a few brief seconds I can see the traces of our presence there—the flattened tundra where the tents had been, lines of crushed plants that were the trails we had made along repeatedly used paths. It is the intrusive geometry of lives lived in a delicate place.

The flight is due south, back to Kangerlussuaq and the airport we flew into after leaving Copenhagen. We are flying at a little over a thousand feet, skimming saddles and ridgetops, grazing unknown surfaces. To the east, the white of the ice cap glares in the sun; it stands nearly a mile above us, forming a relentless horizon that will soon be simply a historical footnote. At times, we fly just a hundred feet above the ice front, close enough to see water gushing from its underbelly in muddy brownish gray rivers that lumber off toward the west, carrying their pulverized lithic load to resting places in a distant sea. On their passage through the land, they choke the rugged valley bottoms, dropping their coarse sand and gravel onto floodplains and valley floors, making new land at the edge of fjords, displacing the blue waters that flow with the flooding tide of oceanic currents.

As we fly south, the clouds eventually vanish and a brilliant blue sky emerges. An endless staccato assault of almost blinding flashes sparkle up from the land—the low-lying morning sun reflecting off of pools and wet surfaces in the water-saturated terrain. I start to hunt for sunglasses, then change my mind—I don't want to be leaving this place and have anything come between me and the experience of it, even though I am in a helicopter flying at twelve hundred feet, the rotor whirling above

our heads at 400 rpm, its incessant *thup-thup-thup* playing in the background.

About halfway to Kangerlussuaq, we pass over a sharp
ridge and see a maze of trails in the tundra of the valley
below us. They are reindeer migration paths, empty and
otherwise featureless. And yet, they embody a history.
They are the ephemeral writings, emblematic of the lives
lived there, the elusive consequences of change and survival in an evolving land.

To our left, the ice pursues its unending effort to dis-
assemble the world's largest island; to our right, beautiful sculpted valleys and sediment-filled fjords finger off
toward the west. The diversity of that scenery under-
scores how inadequate purely analytical descriptions of
natural processes can be.

Suddenly, we cross a saddle in a ridge and see to the
southwest, five miles away and a thousand feet below us,
the concrete and tarmac of the Kangerlussuaq airport—
an engineered construct designed to withstand extreme
seasons.

The helicopter begins to descend and turn. As we fly
in, we see the 767 that will soon jet us across the North
Atlantic. We will be in Copenhagen in time for dinner.

Gently, the helicopter settles onto the tarmac. I
unbuckle my seat belt and climb out, my bare hand resting on the painted aluminum skin of the chopper. Immediately, I am struck by the silky feel; the smooth, polished
surface is like nothing I have touched in the weeks we
have been camping. Although we are standing, more or

less, on the spot from which we had entered the wilderness more than four weeks before, there is no sense that any of this is familiar.

We pull our gear out of the helicopter and toss it in a van—it lands with a hollow metallic clunk. There in that diesel-fueled concrete compound is the essential expression of what we are returning to. My footpath scar across the tundra at our camp seems the definition of nothing.

We are leaving an existence that was attentive to friendship, tides, winds, and the layering of clouds. The new world is one of separation from the natural flow of evolving landscapes and life, a place of borders and boundaries. Even the flat hardness of the tarmac seems odd—the feel of an irregular surface that offers a thousand ways to feel Earth has been intentionally obliterated.

The van takes us across the airfield to the terminal/cafeteria/hotel. We walk into the building, where we check our bags for the flight to Copenhagen. At one end of the hotel are public facilities, available for a small fee. Money, a useless concept back in our little community, has the feel of an odd abstraction. We search for bills we had stashed in some zippered pocket weeks before.

A mild claustrophobia haunts each step as I walk to the shower. After two turns down the rectangular corridor, I am dizzy and disoriented.

Afterward, standing at the sink, preparing to shave off more than a month's worth of beard, the absence of any breeze and the warm humidity of that closed space become even more oppressive. I open a window, which

looks out on the rolling hills that are the eastern termi-
nus of Kangerlussuaq Fjord, and am relieved at the rush
of fresh, cool air.

IMPRESSIONS IV

It was better, I decided, for the emissaries returning from the wilderness . . . to record their marvel,
not to define its meaning. In that way, it would
go echoing on through the minds of men, each
grasping at that beyond out of which miracles
emerge, and which, once defined, ceases to satisfy
the human need for symbols.

—Loren Eiseley

THE SCENE IN MY MIND IS OF THE LEDGE above that cliff
face where the falcon and I met. In front of me, in the
vast abyss of flowing air, swims that river of fish, migrating to their destiny. Existence there is insulated from the
fears that permeate the populated world. What is spoken
comes only from the vanishing voice of wild things in wild
places, the phenomenon of thought an observation point.

We hover in suspension, our thoughts and dreams
bound by the surfaces of those things we know and see.
We are the pioneer species that conceives there is a surface
and something that it cloaks. Scraping shins on bare rock,
bleeding from the touch of raw crystals, walking with saturated boots through thin air, all inform our experience
and create what we see as the natural world. Ice blocks

among the fishes, the shattered, screaming winds blasting against cliff faces, the ooze of juices from seal flesh, the sweet scents from the reproductive organs of flowering life—wilderness becomes, in our presence, the only threshold through which we can freely perceive the significance of our ability to reason, imagine poetry, and create what we cherish as beauty.

Epilogue

Earth is the construct of wandering stardust, accreted from the atomic debris of supernovae and the elemental winds of unknown stars. The gentle fall of interstellar particles, the collisions of comets and meteors and frozen water, gave rise to our planet in a rush of cosmic artistry just over four-and-a-half billion years ago.

The creativity has not ceased; geology and life are the consequence. But to perceive and participate in such richness requires access to an entire spectrum that has become obscured by parking lots and buildings and city streets. To see the content of sunsets and horizons, termites and molecules and life responding to nature through its own creative prowess requires unshaped space. Without wilderness, the essential perspective that allows such seeing is lost.

Our fieldwork and that of other geologists has allowed us to recognize the gross outline of an ancient mountain-building story. To allow the voice of the bedrock to speak unconstrained requires that we look at the details still waiting to be seen at scales we cannot see with the naked eye. That is why we carry hammers and sample bags and labeling pens.

From the samples we collected and shipped back to Denmark, we cut thin slabs and send them to special laboratories where they are glued to glass slides and ground down to the thickness of a human hair. The rock surface is then polished to the sheen of glass. These are "thin sections" through which light can pass, allowing the finest details of texture and form to be viewed and recorded.

Gazing down a microscope at those slender pieces of rock, self-awareness is lost as the mind focuses on fantastical geometries of color and form that no human could have imagined or naked eye perceived. The beauty and fabric of that microscopic realm is a simple expression of the magic of coordinated arrangements of atoms in crystal cages. Hour after hour, as we try to read what has been preserved from the past in those mineral forms, the profundity of the enterprise presents itself. Minerals transforming from one assemblage to another are frozen in geometries that preserve unfolding processes that never reached equilibrium, attesting to the reality that nothing is complete; crystal faces press on enclosing neighbors, annihilating the possibility of empty space; successions of stable arrangements grow over and around one another, delineating changing conditions deep in the earth.

But reconstructing history is more than tabulating sequences. We need a way to obtain an age for when a mineral formed or a texture developed. For rocks that have experienced more than three billion years of history,

View through a microscope.

constructing that scaffolding of time requires a recording machine with a robust memory. Fortunately, the mineral zircon performs that function well.

Zircon is a mineral mainly composed of zirconium, silicon and oxygen. It is a resilient phase, stable at most temperatures and pressures experienced by rocks in the middle and deeper levels of Earth's crust. It, too, is tough—it can travel tens or hundreds of miles in streambeds, pounded on and scraped by boulders and cobbles, resisting abrasion through the uniquely powerful bonds of its crystal lattice.

Because of its unique combination and arrangement of the atoms that determine its structure, zircon also readily accepts traces of uranium, which virtually all rocks possess. This simple fact makes it one of the most profoundly important minerals for reconstructing the history of the crust, especially in places where tectonic activity has heated and compressed the rock under multiple extreme conditions. The uranium in zircon slowly decays radioactively at a constant rate, disintegrating into lead, thorium, and helium that accumulate over time. Hence, the ages of rocks can be determined if the concentrations of these elements are measured—zircon as geological clock.

To obtain the zircon for dating, part of the sample must be crushed and sieved until the minute zircon crystals can be separated out. A suite of the grains is then mounted on an epoxy disk, polished so the interiors of the grains are exposed, and then analyzed. Inevitably, complexities arise. When the zircon crystals are viewed at high magnification, it is immediately obvious that they are often not homogeneous. They commonly contain zones that look much like tree rings, which surround the central interior core of the crystal—a record of changing conditions in which new zircon grew over old. Often the thickness of these growth bands is only a few millionths of an inch or less—these currently remain unreadable and cannot be dated.

But even though we are not capable of resolving the finest growth rings in zircons, our analytical techniques applied to broader bands allowed us to obtain hundreds of individual dates, making it possible to construct in more detail than ever before the evolution recorded in the landscape's bedrock.

SOME OF THE SAMPLES WE SELECTED for dating were from the flowing, plastic forms we found at the eastern end of Tunertoq Island. Poring over the data with colleagues, we discovered how unexpectedly old some of the rocks were. Cores of numerous zircons turned out to be nearly 3,400 million years old, older than virtually the entire terrain on the other side of the shear zone. This implied that an ancient continental body extended farther north but not to the south. As such, those rocks defined one boundary of the vanished ocean.

Around those old cores were younger zircon rings, many dating an event about 2,750 million years ago. That age corresponds to a major convulsion seen in old continents around the world, as well in the rocks south of the shear zone. The significance of this remains unclear, although it seems to indicate a time at which much of the mass of continents on the planet bubbled out of the mantle. Such evidence establishes a correspondence between what was occuring in this terrain and processes happening elsewhere—a common denominator demonstrating that this part of Greenland is typical continental crust.

Cutting through these old rocks was a dike with homogeneous zircons that formed 1,805 million years ago, which turned out to be a defining time for the region—the time in which the continents were colliding.

Farther to the east, the rocks forming the huge igneous massif recognized by Kalsbeek and his coworkers in 1987 are now well dated, based on zircons we collected as well

as those of a similar age obtained by a few other researchers. The measured ages prove there was active plate movement and Andes-like volcanism throughout that region
between 1,875 and 1,980 million years ago.

An active volcanic system that lasts for 100 million
years helps constrain, through simple arithmetic, the size
of the ocean that must have been consumed. We know
that today the rate at which ocean crust descends into
subduction zones is usually between one and five inches
per year. If we assume a low rate for the convergence of the
plates back then, the amount of crust consumed would be
about three thousand miles, which is almost the distance
between New York City and Lisbon, Portugal. In other
words, the ocean that the pillow basalts floored may have
been about the size of the North Atlantic today.

But is the age of the pillow basalts consistent with being
part of an ocean basin that was active between 1,875 and
1,980 million years ago? Using the same analytical techniques, we dated the pillow basalts and found them to be
at least 1,895 million years old, which shows they likely
were part of that long-vanished ocean floor.

Other samples collected in the shear zone where deformation and metamorphism were concentrated provided a
variety of ages, but all of them within the range of 1,720
to 1,820 million years ago. This 100-million-year duration
for the complete mountain building cycle is consistent with
what is known for similar types of collisional mountain-
belts, such as the Himalayas and Alps, both of which are
still active and have millions of years to go before they

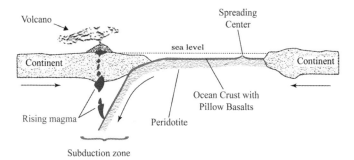

Before the collision. At about 1,890 million years ago, continents that were involved in the collision, as well as the subduction zone and volcanic system that were the main active plate tectonic elements at the time, would have been arranged, schematically, as shown. The ultrahigh-pressure (UHP) region would have been just below the area where the rising magma bodies are coming off of the down-going pillow basalts and peridotites, and the high-pressure (HP) rocks would have been above that region.

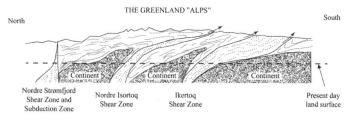

A new interpretation. A schematic drawing of a cross section through the mountain system that formed when the collision of continents had nearly finished, about 1,720 million years ago. The arrows indicate movement along major faults that were rooted in the shear zones we now recognize. The darkly shaded continents would have been joined together before the collision. The oldest continental rocks that compose the northern continent are to the left of the Nordre Strømfjord shear zone. Metasediments and other rocks are indicated by the thin wavy and folded lines. This figure is a modified version of a model originally drawn by Kai Sørensen.

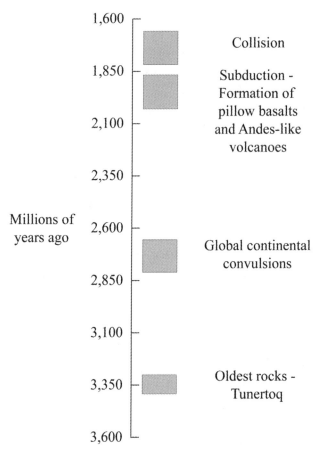

A time line of the most significant events preserved in West Greenland geology. The ages are mainly based on dates obtained from zircons. The length of each bar encompasses the majority of ages that define each event. The activation of the Nordre Strøm-fjord shear zone (NSSZ) took place during the last few million years of the collision. For reference, the earth formed about 4,560 million years ago, and the oldest preserved continental fragments on earth are about 4,100 million years old.

cease. The Himalayan system began forming sixty million years ago, the Alps are at least thirty million years old.

The ability to date the mineral grains in our samples provides a chronology upon which the journey of a rock can be detailed. Using that approach, a three-dimensional model through time can be constructed for the history of the terrain.

The singed-hair rock that we examined with microscopes and discovered was filled with garnets and olivines and spinels contained a startling history of burial at a depth of at least forty miles, an HP metamorphic environment. Up to that time, none of us had imagined that any of the rocks in this region had traveled more than fifteen miles down. We wrote reports and published papers and looked at more samples in the basement archives of Aarhus University, seeking confirmation that such rocks were not enigmatic anomalies.

The UHP samples were found during those explorations in the archives. For months, we examined thousands of samples that had been collected over decades by a small cohort of faculty and students working on master's and doctoral degrees on Greenland geology. Out of all those samples, we found two that preserved evidence of the same very deep burial. The samples came from sites tens of miles farther to the west of where we had been working, but along the same belt of unusual rocks, and along the northern edge of the Nordre Strømfjord shear zone. The samples from both of those sites had identical characteristics. One sample, ironically, had been collected by Kai when he and Fleming Mengel, a student of his, had worked in the region nearly forty years before. Kai didn't remember collecting it. The other sample came from a site near Giesecke Sø and had been studied in the late 1960s by Steen Platou, who was working on a graduate degree at the time. Those samples became the core of a small collection that proved that fragments of the region had, indeed, been pushed to extraordinary pressures, surviving a round-trip circuit to depths greater than 150 miles. They have become the oldest known samples that preserve evidence of travel into a deep subduction zone, the place where the ocean floor descends hundreds of miles into the mantle when tectonic plates meet. Prior to these discoveries, no direct evidence existed that such plate tectonic-driven processes occurred any further back in time than 900 million years ago. These samples pushed that age limit back to at least 2,000 million years.

AT THE TIME WE DISCOVERED THE SAMPLE from Steen Platou's field area, he was a retired farmer, living on the outskirts of Århus, Denmark. We visited him at his farm-house, went over his notes and maps, and talked to him about what he remembered of the place. Eventually, we all realized the only smart thing to do was go with him to his old field area. So, in the summer of 2012, we returned to the place Steen had explored on his own, an area no one had visited since he last worked there in 1969. He was now in his early seventies. We spent days slowly walking through the terrain, with him as our guide. He smiled often, smoking his pipe, obviously enjoying the return to his old haunts. One afternoon toward the end of our stay, he proudly pulled up his shirt to show us that his belt no longer fit—he had lost so much weight hiking miles in that wilderness that it was too big and there were no holes left for the buckle clasp.

Steen died a few months after his trip with us. He was happily working on the maps we needed to assemble our data when he suffered a stroke. The samples he had long ago collected, and the ones we collected with him on our last trip, formed the core of evidence documenting a unique history. The data and samples thoroughly supported the argument of Kalsbeek and his coworkers that there once had been a volcanic system on top of a Greenlandic version of the Andes.

We found the suture zone that demarcated the boundary between continents and the remnant pieces of the ocean floor that had once separated those landmasses. Through that work, and related studies by others, the Nagssugtoqidian shear zone is unequivocally established as the last major deformation at the end of the grinding collision between ancient continents, a fault system much like those active systems emanating from the Himalayas today. And contained within this system are the rare vestiges of rocks carried 150 miles down into the mantle and returned to the surface. They are the oldest known record of an entire terrain on the surface of the world that had descended to such depths—the exposed remnant of the earliest known plate tectonics and subduction. It was Steen who had found them.

WE RIDE THE BACK OF A TIDAL FLOW, coursing among the rocks along the shore, which is glittering in the rainbow light of comb jellies. John turns us farther out into the current. The gneisses and schists, indeed every lithic form cradling the flow, sing of a past we revel in. In the wake of our boat is the shape of the future. The flexing pontoons respond to each crest and trough, slowing, accelerating, a brief sideways shove, a chance to splay water.

Since our last expedition, polar bears have been sighted in our field area in the summer—a place they never were before. To satisfy the needs of our funding agencies, we would have to carry rifles for our protection if we were to return when they are there.

In my mind I again see the dissolving landscape of tundra tufts resting on boulders in that encroaching bay from years ago. I watch reindeer bones decay, ice melt, new surfaces appear. Despite its inevitable dissolution and change, what wildness remains will forever quietly, irresistibly beckon.

SETTLEMENTS AT THE EDGE OF WILD PLACES are the punctuation points in nature's wilderness stories. By providing a human element, they shape emotion and responses, affecting the texture of place. Because they are at the edge of what is inhabitable, they define what it means to exist in harmony with untouched landscapes; the wisdom contained in such places is deep.

Early one morning, John and I walk the streets of Aasiaat, looking for the house of one of the people who will help with our logistics. He is Inuit, an older resident whose home is on a small knoll overlooking Disko Bay. Draped on the roof is the drying pelt of a reindeer he recently killed; hanging from the second-floor window frames of the modest house are strips of curing reindeer meat. Huskies are tied out back to their respective little houses. The sled they will pull in the coming winter stands upright next to them, its gracefully curving white runners arcing toward the sky.

As we approach the front door, a strange sound moves through the air. Multitoned, the pitch slowly changing from low to high and back again, the song roils up from the bay. I turn and look out over the ice-studded water, but all I see is white-speckled blue, a shimmering, placid reflection of sky. Just as we get to the front porch, the bay surface erupts in three massive waves, out of which the huge gaping mouths of humpback whales slowly rise. The sonorous noise has stopped, replaced by the sound of rushing water draining through baleen. They are feeding, the song the mechanism they use to herd and concentrate the small sea life they live on.

When we finish our business, we head back to the seamen's hotel we are staying at to meet up with Kai. The road we walk goes by the beach that runs along the small harbor.

We stop at a small cluster of white canvas booths where fish and seal meat are sold by a few of the local fishing people. As we look over the flatfish, fjord cod, Arctic char, and some fish I do not know, a small outboard roars into the harbor, throttles down as it approaches the beach, then slowly glides onto the sand. A large man in yellow chest-high waders steps out of the boat and hauls out long, thick strips of deep red seal meat. We watch him bring them to one of the booths and negotiate with the Inuit woman standing behind a table. After a brief exchange, she makes room for his goods among the display of fish she has laid out. He heads back to the little boat and returns with slabs of blubber. Once he has been paid, he walks back to the skiff, pushes out into the bay, and pulls the chord on the outboard engine. As it roars to life, he, standing, slowly maneuvers it through the moored boats in the harbor, then opens the throttle and roars away, disappearing behind a headland.

This is a traditional scene, an old way of commerce and sustainable coexistence with wildlife and wild land that has changed little over hundreds of years. But the catch of cod is diminishing, whales are harder to find, the migratory routes of the reindeer are more difficult to discover, and the seal population is now out of balance with its ecosystem. What was once a strenuous but consistent existence is now threatened.[*]

But this is not a situation unique to Greenland; on

---

[*] F. Karlsen, 2009. Management and Utilization of Seals in Greenland. The Greenland Home Rule Department of Fisheries, Hunting and Agriculture. 28 pages.

every continent, wilderness is being consumed, and the people who have depended on it, living at its fringe and within its embrace, are forced to relinquish what they cherish. With infinite hubris, the modern world is imposing the consequences of its industrial avarice on lifestyles it knows nothing of. The moral bankruptcy of the rationalizations for the destruction of wilderness and the people who live in harmony with it is staggering. That many are, in fact, angry and seeking ways to mitigate impacts is heartening, but the pushback is formidable. The moral outrage we all should feel seems meager against the economic juggernaut.

Compounding the consequences of this economic monstrosity is the diminished role wilderness has in our everyday lives. Seldom is it mentioned on the news, rarely is it considered in politics, and its presence on social media is almost nil. In his profoundly influential "Wilderness Letter" in 1960, Wallace Stegner wrote:

> [Wilderness] is good for us when we are young, because of the incomparable sanity it can bring briefly, as vacation and rest, into our insane lives. It is important to us when we are old simply because it is there—important, that is, simply as an idea.

The message of this letter is withering, but its urgency is more powerful today than ever.

Humanity is grounded in community, which requires cooperation and shared experiences. As politics and economic self-interests bulldoze through the world and wilderness recedes, we risk losing access to our own essential wildness. Whether through direct experience or through poetry, art, or song, we must share and celebrate the wild so that it may be saved. The lives lived there—of all species—are worthy of our recognition and respect, the land, our awe, art, and dreams.

Glossary

anorthosite: A rock formed from magma that is composed of small amounts of orthopyroxene and mainly contains plagioclase, a mineral rich in calcium, sodium, aluminum, and silicon. It is believed to be a common rock at the bottom of continents.

Bugt: A Danish word for an ocean bay.

country rock: The main bulk of lithologies that compose a terrain. Also used to describe the rocks into which magmas invade.

defile: A valley or pass in a terrain with significant topographic features.

fjord: An inlet, often with high, sheer bounding walls, formed where a glaciated valley is invaded by the sea.

foehn: A strong, warm wind that forms on the downslope side of a topographic feature. Originally applied to meteorological phenomenon in the Alps, the name is now also applied to winds that form on the downslope side of major ice sheets, such as the Greenland ice cap.

fractionate, fractionation: A process of separation. In scientific applications, the term is usually applied to situations in which one material, such as a solid or gas, separates out of another material, such as a liquid.

gneiss: A metamorphic rock that has experienced high temperatures and pressures and contains layers of differing mineralogy. A gneiss usually displays banding seen as different-colored layers. A gneiss can be formed from virtually any kind of rock (igneous, sedimentary, metamorphic) that is sufficiently heated and sheared.

isotope: Two or more forms of a specific chemical element in which the nuclei contain the same number of protons but different numbers of neutrons. Some isotopes are stable, while others are unstable and radioactively decay to other elements.

lee: The downwind side of a sailing vessel or that portion of a shoreline, landmass, or object that is protected from wind, as opposed to the windward side which is directly confronted by the wind.

lithic: Composed of stone

orthopyroxene: A mineral that forms at high temperatures in certain igneous and metamorphic rocks. It is mainly composed of iron, magnesium and silicon.

palsa: A geological term for a rounded mound of soil a few feet across rising out of a moist or watery region. The form of the mound is determined by a frozen ice core several to tens of feet below the surface of the palsa.

pingo: A geological term for a large version of a palsa, sometimes reaching hundreds of feet in diameter.

protolith: The precursor rock from which later rocks formed through metamorphosis. Often, being able to identify a protolith is a powerful way to reconstruct an earlier environment or setting.

relict: A feature, artifact, or form that has survived from some earlier time.

schist: A metamorphic rock with paper-thin laminations and layers formed by platy or elongate minerals.

sillimanite: A white metamorphic mineral that tends to occur in elongate, needlelike forms. Its presence usually signifies that clays or other aluminum-rich materials were present in the rocks that were metamorphosed.

stope, stoping: A process in which overlying materials are detached and engulfed by an upwardly rising magma body. The term is usually used in the mining industry when overlying material in a mine is removed, but it has also been applied to processes involving the movement of molten rock (magma) up through the crust.

subduction: The process in which one tectonic plate descends beneath another.

tectonic plate: An expanse of crust and upper mantle that slowly migrates over the Earth's surface. There are eight major tectonic plates and numerous smaller plates on the Earth's surface. The individual plates are relatively rigid and it is for this reason that mountain systems form when plates collide with each other.

tundra: A cold, treeless region that occurs at high latitudes or elevations. The growing season is short. The combination of weather conditions results in a unique biome of plants.

twin: A crystal structure in which the orientation of the lattice of atoms composing a crystal is oriented in a way that is different from that of adjacent parts of the crystal.

ultramafic: A rock type that is rich in iron and magnesium and low in silica, aluminum, sodium, and potassium. Ultramafic rocks make up the bulk of Earth's volume, occurring as the predominate rock type in the mantle.

Acknowledgments

ATTENTION TO WILDERNESS HAS BEEN VOICED for centuries, the literature rich with diverse visions and personal experiences, each grounded differently. Thank you to a few of those who have exposed the necessity of reflection on wilderness and being, and who inspired humility in the process, in no particular order and woefully incomplete:

Loren Eiseley—*The Immense Journey* (1957)
Ilya Prigogne—*From Being to Becoming* (1980)
Freeman Dyson—*Disturbing the Universe* (1979)
Henry David Thoreau—*Walden* (1854)
John Muir—*The Mountains of California* (1875), *My First Summer in the Sierra* (1911), and *The Yosemite* (1912)
Aldo Leopold—*A Sand County Almanac* (1949)
Edward Abbey—*Desert Solitaire* (1968), and *The Monkey Wrench Gang* (1975)
Robert MacFarlane—*The Wild Places* (2007)
Margaret Mead—*Coming of Age in Samoa* (1928)
Rachel Carson—*Silent Spring* (1962)
Gontran de Poncins—*Kabloona: Among the Inuit* (1941)
Peter Matthiesen—*The Snow Leopard* (1978)

Gary Snyder—*Riprap and Cold Mountain Poems* (1959), *Turtle Island* (1974), and *The Practice of the Wild* (1990)

Barry Lopez—*Arctic Dreams* (1986)

Rockwell Kent—The first to paint Greenland for a Western audience

Wallace Stegner—*Angle of Repose* (1971)

John Steinbeck—*The Log from the Sea of Cortez* (1951)

Henry Beston—*The Outermost House* (1928)

E. O. Wilson—*Consilience* (1998)

Annie Dillard—*Pilgrim at Tinker Creek* (1974) and *Teaching a Stone to Talk* (1982)

Gretel Ehrlich—*The Solace of Open Spaces* (1985), *Islands, the Universe, Home* (1991), and *This Cold Heaven* (2001)

Elsa Marley—*Blue Ice Series* (2009) and other magnificent paintings.

Terry Tempest Williams—*Refuge* (1992), *When Women Were Birds* (2012), and *The Hour of Land* (2016)

Thank you to Kai and John, who began the Greenland adventure, invited me to be part of it with them so many years ago, and sustained the passion for life and place that allowed Team Alpha to materialize. Their enthusiasm, heart, and honesty have served us and the science they practice, well. To the people of Greenland, who have sustained a culture that deeply and intimately acknowledges and respects the wonder and power of the wild world within which they thrive. Their struggle to persist in spite of the pressures imposed on them from external forces should inspire each of us to be more than what we have chosen to be. To Lucia Milburn, Peter Seitel, and John Winter for companionship in the field during my first expedition.

Deep appreciation to Katharine Turok, whose thorough, insightful and sensitive editorial reviews transformed a manuscript into a book. Her patience and grace in walking a naïf through a small portion of the vast landscape of writing has been boundless. Thank you to Dawn Raffel, who provided guidance that nurtured the book and helped it grow. To Erika Goldman, whose patient and tireless editorial efforts took it to maturity, I am forever grateful. To Carol Edwards, thank you for an extraordinary effort at clarifying and refining the intent of the text. Thank you to my agent, Malaga Baldi, who persevered and encouraged, finding a home for this work. And, thank you to Elana Rosenthal and Molly Mikolowski, for their dedicated attention and perceptive vision.

To Carolyn Feakes, I offer my deepest heartfelt gratitude for boundless patience as, day after day, she put up with the endless effort to find what needed to be said. To Sabina Thomas, Martha Hickman Hild, Annemarie Meike, Lucia Milburn and Dirk Sigler for their generous offerings of time and insight and commenting over the years on various versions of this evolving book. Thank you to Lawrence Millman, for providing mycological insight. And thank you to the faculty and students at College of the Atlantic, Bar Harbor, Maine, for rich engagement in discussions about wilderness and its values.

Funding for the research we have conducted in

Greenland over the years has been provided at various times by the U.S. National Science Foundation, the Danish Research Council, the Greenland Geological Survey (GGU) and the Geological Survey of Denmark and Greenland (GEUS). The support of these organizations is gratefully acknowledged.

Quotation Source Notes

Page

9   Katherine Larson, "Solarium," *Radial Symmetry* (Yale University Press, 2011).

     Alan Watts, *Cloud-Hidden, Whereabouts Unknown: A Mountain Journal* (Vintage Books, 1974).

31   George Bancroft, *The Necessity, the Reality, and the Promise of the Progress of the Human Race: Oration Delivered Before the New York Historical Society, November 20, 1854* (New York, 1854).

33   John Steinbeck, *The Log from the Sea of Cortez* (Penguin Classics, 1951).

105 Barry Lopez, *Arctic Dreams* (Scribner, 1986).

107 John Muir, *My First Summer in the Sierra* (Houghton Mifflin, 1911).

147 Alfred Lord Tennyson, Canto 123, *In Memoriam A.H.H.* (London, 1850).

149 Annie Dillard, *Teaching a Stone to Talk* (HarperCollins, 1982).

197 Loren Eiseley, *The Immense Journey: An Imaginative Naturalist Explores the Mysteries of Man and Nature* (Random House, 1957)

BELLEVUE LITERARY PRESS is devoted to publishing
literary fiction and nonfiction at the intersection of
the arts and sciences because we believe that science and the
humanities are natural companions for understanding the
human experience. With each book we publish, our goal is
to foster a rich, interdisciplinary dialogue that will forge new
tools for thinking and engaging with the world.

To support our press and its mission, and for our full
catalogue of published titles, please visit us at blpress.org.

BELLEVUE LITERARY PRESS
New York